THE
PURSUIT OF
PARENTHOOD

THE

REPRODUCTIVE TECHNOLOGY

PURSUIT OF

FROM TEST-TUBE BABIES

PARENTHOOD

TO UTERUS TRANSPLANTS

Margaret Marsh and Wanda Ronner

JOHNS HOPKINS UNIVERSITY PRESS

BALTIMORE

© 2019 Johns Hopkins University Press
All rights reserved. Published 2019
Printed in the United States of America on acid-free paper
9 8 7 6 5 4 3 2 1

Johns Hopkins University Press
2715 North Charles Street
Baltimore, Maryland 21218-4363
www.press.jhu.edu

Library of Congress Cataloging-in-Publication Data

Names: Marsh, Margaret S., 1945– author. | Ronner, Wanda, author.
Title: The pursuit of parenthood : reproductive technology from test-tube babies to uterus transplants /
    Margaret Marsh and Wanda Ronner.
Description: Baltimore : Johns Hopkins University Press, [2019] | Includes bibliographical references
    and index.
Identifiers: LCCN 2018048633 | ISBN 9781421429847 (hardcover : alk. paper) | ISBN 1421429845
    (hardcover : alk. paper) | ISBN 9781421429854 (electronic) | ISBN 1421429853 (electronic)
Subjects: | MESH: Fertilization in Vitro—history | Reproductive Techniques, Assisted—history |
    Infertility—history | Infertility—therapy | Reproductive Medicine—history | Health Policy |
    United States
Classification: LCC RG135 | NLM WQ 11 AA1 | DDC 618.1/780599—dc23
LC record available at https://lccn.loc.gov/2018048633

A catalog record for this book is available from the British Library.

Special discounts are available for bulk purchases of this book. For more information, please contact
Special Sales at 410-516-6936 or specialsales@press.jhu.edu.

Johns Hopkins University Press uses environmentally friendly book materials, including recycled
text paper that is composed of at least 30 percent post-consumer waste, whenever possible.

For our dear friends

Maryellen and John Alviti,

who have shared so much

of our lives

CONTENTS

Preface   ix

Introduction: The Past as Prologue   1

1 Test-Tube Babies Just around the Corner   11

2 From First Dream to First Baby   31

3 IVF Comes to America   53

4 From Miracle Births to Medical Mainstream   78

5 The Elusive Search for National Consensus   102

6 A Lot of Money Being Made   121

7 Beyond Infertility   146

8 Can the Wild West of Reproductive Medicine Be Tamed?   184

Appendix: Assisted Reproductive Technologies
by (Some of) the Numbers   213

Acknowledgments   221

Notes   227

Index   265

■

PREFACE

People have been known to introduce us at conferences as the infertility sisters. Margaret is a historian of women with a special interest in gender and the family; Wanda is a gynecologist. For nearly three decades now, the two of us have been thinking, talking, and writing about the history of infertility, reproductive medicine, and reproductive sexuality. And we really are sisters. We never imagined that our collaboration would span our careers when we started out, oh so casually, in 1988. Wanda had just returned to Philadelphia to practice after she completed her residency. At the time, Margaret was close to finishing her second book, and Wanda said, "Why don't we do a little project in the history of gynecology? Maybe a talk, or a paper?" We had always been interested in each other's work, and we both thought it would be interesting to merge our two disciplines into a joint project. That "little project" ended up becoming *The Empty Cradle: Infertility in America from Colonial Times to the Present*. That first book led before long to our second, *The Fertility Doctor: John Rock and the Reproductive Revolution*, which explored the development of the field of reproductive medicine through the life and career of one of the most prominent infertility specialists of the mid-twentieth century, a man who later became famous as the co-developer of the oral contraceptive.

After spending so many years thinking about these subjects, how could we not want to bring our unique perspective to the late twentieth-century transformation in reproductive medicine and technology? In vitro fertilization—the creation of embryos outside a woman's body—did more than allow women with certain diagnoses of infertility to conceive. It also made possible the decoupling of pregnancy and genetic parenthood. When Louise Brown was born in 1978, the press called her the world's first "test-tube baby." Because she had actually been conceived in a petri dish, not a test tube, the medical profession sought to shift the terminology to in vitro fertilization, usually called by its initials: IVF. That name caught on, but the test-tube moniker never entirely disappeared. Today, the medical profession calls IVF and other procedures where eggs or embryos are handled outside the body "assisted reproductive technologies," often abbreviated ART.

These new technologies do not exist in some medical vacuum but reach to the core of what it means to be a family. In part because of what they can do, and in part because of the way the field has been developed in the United States, their development and use have been controversial from the start, raising numerous questions. Who gets to decide which groups can have access to reproductive services? Is infertility a disease and if so, who should pay for its treatment? Should there be limits to the lengths to which couples and individuals can pursue assisted reproduction to make their idea of family a reality? Should decisions about whether and how to reproduce be made by individuals or governed by societal norms? And if we leave the decision up to individuals or couples, can we do so without maintaining the kinds of long-standing racial and economic disparities in care that make it possible for some people to have children while others go without? These are some of the questions that we seek to answer. In the final volume in what the family calls our infertility trilogy, we explore the contemporary technological pursuit of parenthood, beginning with in vitro fertilization and ending with such new developments as mitochondrial replacement techniques and uterus transplants.

In all our work, our key interest has been in the links among medical science and practice, the experiences of patients, and the cultural and political frameworks within which the interactions between people and their doctors take place. In this book, we tell the interconnected stories of the scientists and physicians who developed and employed these technologies and the women and men who have utilized them. We examine the larger social and political frameworks in which they were developed and explore the moral and ethical controversies they engendered.

As we've been working on this book, we have talked, formally and informally, with some of the doctors who were responsible for the first few cohorts of IVF babies. Several of these men and women are still practicing in the field, while others have moved on. There were multiple paths to their pioneering roles in IVF, some of them more surprising than others. We've also been moved by the stories we heard in unexpected places from women and men who experienced these technologies. At conferences, at social events—almost everywhere, it seems—we meet someone who has had a baby after IVF or has failed to become pregnant, is thinking about trying IVF or knows someone who just did, or is on their third round of treatment and wondering whether to keep at it, and if they stop, what then? We were at a conference a few years ago where one of the speakers happened to mention, almost casually, that she had altruistically donated eggs to her best friend many years ago.

To date, around eight million babies have been born as a result of assisted reproductive technologies, about one million of them in the United States. Here we tell the story of this phenomenon and assess its impact, not just on those who developed or used the technologies but on all the rest of us who have heard about or encountered them.

THE
PURSUIT OF
PARENTHOOD

INTRODUCTION

THE PAST AS PROLOGUE

Wendy had begun to think she would never have the children she had always wanted. She was twenty-eight, after all, and she had already tried the usual infertility treatments, including medication to help her produce extra eggs and intrauterine insemination with her husband's sperm.[1] But then, in September of 1983, the University of Pennsylvania announced the first birth in its year-and-a-half-old in vitro fertilization (IVF) program. It turned out that she was not out of options after all. Wendy and her husband decided to give their dream of parenthood one more try with this new technology. Her first step involved a diagnostic laparoscopy to determine whether her ovaries were accessible. She was given general anesthesia, then her surgeon made a small incision beneath her navel and used carbon dioxide to distend the abdomen, after which he inserted the laparoscope, a thin, lighted telescope he would use to view the pelvic organs. The laparoscopy showed that her ovaries were visible and accessible, which meant that it would be possible to retrieve her eggs by this means. Fertilizing those eggs would require a second laparoscopy, after Wendy's ovaries had been stimulated to produce multiple eggs. Wendy injected herself daily with Pergonal, which at the time was the most effective medication for that purpose.[2] She then had the second laparoscopy, and her eggs were retrieved and fertilized with her husband's sperm.

The couple ended up with five embryos. At the time, freezing leftover embryos—called cryopreservation—had not become a reliable option, so all five embryos were transferred into Wendy's uterus using a small catheter. Wendy conceived on her first try, and she was elated to discover that she was

■

1

carrying twins. What neither she nor her doctor knew was that a third embryo had implanted in one of her fallopian tubes. When Wendy was around eight weeks pregnant, excruciating abdominal pain sent her to the hospital. The tube had ruptured. She remembers being told that if the doctors had waited another hour to operate, she might have died. Wendy did not lose the twins, but she later went into premature labor, and for six months, as her doctor told reporters after her babies were born, she was in and out of the hospital.[3] Wendy said that she stayed in bed for long stretches of time, forbidden to get up and afraid of doing anything that might cause her to miscarry. When she was thirty-two weeks pregnant, she gave birth to twin boys. One weighed 3 pounds, 12 ounces, and the other was even smaller, at 2 pounds, 7 ounces. It was all worth it, she recalled. "I felt special [to be able] to get the kids I always wanted," she told us in 2018, when her sons were almost thirty-four years old. "I needed to be a mom."

Wendy's story is harrowing, with the ectopic pregnancy, followed by anxiety over whether the surgery to remove her tube would cause her to lose the twins, and the months in and out of the hospital, on bed rest, until the twins were born. Her experience was not the norm, even in 1984, when so much about the new technology was still unknown. In other ways, however, Wendy was a typical patient of that era. She was young, she was married, and her husband was fertile. She had been unable to conceive naturally, as her doctor explained, because "a number of factors, including damaged fallopian tubes, prevented the sperm from reaching the egg." But if she was a typical patient of the time in most ways, she was unusual in one important respect—Wendy was among the few for whom the technology worked. Steven Sondheimer, the reproductive endocrinologist who treated her, said that at the time, "it was like a miracle" every time one of the IVF patients became pregnant. Wendy felt the same way. "I feel like I am holding two miracles," she told reporters from her hospital bed after the boys were born.[4]

Wendy was among the first group of women for whom the birth of a child was made possible by what is now called assisted reproductive technology, or ART, which refers to techniques that involve the handling of eggs and embryos outside of a woman's body to help her become pregnant.[5] In the 1980s, when Wendy became pregnant, the language was different. Doctors used the term "in vitro fertilization," while the press and the public were more likely to talk about "test-tube babies." Today, IVF is no longer new, and although it remains the most common form of ART, it is often used with a range of additional procedures. The current terminology is designed to capture the full range of the processes by which conception begins outside the human body.

In the more than three decades since Wendy's twins were born, helping couples and individuals have children with these new technologies has become a substantial medical enterprise. Both the technologies and the range of women and men who use them have expanded dramatically, as Jean's story illustrates. "I wasn't one of those 'have to have kids' kind of people," she was saying as she bounced her eleven-month-old baby on her lap. Jean is forty-four and has been married for eight years. She had her baby in 2017.[6] Jean and her husband Kevin had been a couple for more than fifteen years when they married; until then they had just never found a good time to settle down. Her job, or his job, always seemed to keep them in different parts of the country. Even after they got married, she said, they weren't thinking about children, except in the abstract and at some future time. Jean was about thirty-eight years old when the couple decided to stop using birth control and she started tracking her ovulation cycle. It was easy, she remembered, to do it electronically. "They have an app," she said, "for everything." Three years or so went by without a pregnancy, and now Jean began to be concerned. She asked her gynecologist to recommend a specialist and was referred to a fertility center, where she and her husband saw a reproductive endocrinologist, a gynecologist with fellowship training in the diagnosis and treatment of infertility and expertise in assisted reproductive technologies.

Wendy's first step toward pregnancy had been surgery. Jean's was blood work. Today, before deciding on a treatment plan, fertility specialists order a series of blood tests that monitor a woman's hormone levels, along with another blood test and an ultrasound to determine whether her eggs are fertilizable. After that, Jean had a hysterosalpingogram, a specialized x-ray examination of the uterus and tubes that can evaluate both uterine and tubal factors that cause infertility. Meanwhile, her husband's semen was analyzed to determine the quantity and quality of his sperm. In other words, Jean and Kevin had a routine infertility evaluation. Depending on the results, many couples can be treated without having to turn to IVF.[7] But because Jean was forty-two and her husband's sperm count was low, they decided not to try any of the medical treatments but to move immediately to IVF. The infertility specialist, Jean remembered, wanting to make sure that they were not misled, told them that their chances of succeeding were only around 4 to 10 percent.

Those numbers were discouraging, but during her first cycle, after inject-ing herself with gonadotropins (the modern-day successors to the Pergonal that Wendy used), Jean produced thirteen eggs, a surprisingly large number for a woman of forty-two. The doctor performed ultrasound-guided egg re-trieval, for which Jean had light sedation, to remove the eggs. They were then fertilized using a procedure called intracytoplasmic sperm injection, in which a single sperm is injected directly into the egg. Introduced in 1992, ICSI was the first truly successful treatment for male infertility. It is also used when prior IVF cycles have failed, for unexplained infertility, and for variable sperm counts.[8]

Because of her age, Jean said, she and Kevin also elected to use a technique called preimplantation genetic screening, or PGS, which involves removing a few cells from each embryo and having them analyzed for chromosome abnor-malities. To the couple's great disappointment, only one of the embryos "came back good." That embryo was frozen while the couple underwent a second cy-cle. This time, every single one of their embryos were found to have abnormal chromosomes. After two retrieval cycles, thirty-four eggs and just one "good embryo"—to use Jean's words—they decided to have it transferred. The trans-fer worked. Jean conceived and enjoyed an uneventful pregnancy and delivery. She was so grateful, she said, and she knew she was lucky. Some of the "other moms going through it," she recalled, "were having struggles." She "felt bad talking" to them, she told us, "because everything had gone good for me."

The experiences of Wendy and Jean, at one level so different, were similar in others. Both women were married and using the sperm of their husbands to conceive. They and their husbands were college-educated professionals with the financial means to afford what was then—and now—an expensive proce-dure. In the 1980s, virtually all the women who conceived with IVF were mar-ried, most were in their twenties or early thirties, and their husbands were providing the sperm. Today, the age range is broader—about 22 percent of the women undergoing IVF are over age forty, like Jean. Same-sex couples and unmarried women are also among those using assisted reproduction; never-theless, the vast majority of egg retrieval cycles done in this country today in-volve heterosexual couples in their twenties and thirties seeking a pregnancy using their own sperm and eggs. Finally, both Wendy and Jean, and their hus-bands, are white, as are most Americans using assisted reproduction.

Some observers have argued that stratification by race and class was built into the technology right from the beginning, and these stories may confirm that view. It might have been different. A planned but never implemented trial of IVF in 1978 enrolled both African American and white couples. And in Southern California in 1981 and 1982, an IVF program directed by a young doctor named Richard Marrs included women who were broadly representative of the region's population by class and race. The first birth among his patients was to a Korean American woman of modest means. But by the middle of the 1980s, most couples seeking IVF were white, and since then, entrenched social and cultural attitudes have contributed to significant racial and socioeconomic disparities in access to these procedures.

Neither Wendy nor Jean was thinking about larger societal implications when their babies were conceived in a petri dish. For them and their husbands, IVF was a personal and private decision. It may have been an unconventional road, but it led to a conventional end. If Wendy had known, as she was lying in the hospital fighting to hold on to the pregnancy that this new technology had made possible, that IVF was the subject of heated political and moral debates, it would have made no difference. All she wanted was to have her babies. And if Jean had been told that preimplantation genetic screening had engendered criticism on ethical grounds, it is unlikely she would have changed her mind about using it. She knew that at her age there was a greater risk of chromosome abnormalities such as Down syndrome, and PGS gave her peace of mind.

Medicine and Morality, or Politics and Policy?

Wendy and Jean were making personal reproductive choices when they decided to use IVF. And so were all the other couples and individuals who have been treated with assisted reproductive technology over the course of the nearly forty years since the first IVF program was opened in the United States in 1980. Their decisions, however, have not taken place in a social or political vacuum. IVF was controversial when Wendy was pregnant with her twins, but today, married couples undergoing the procedure with their own eggs and sperm barely raise an eyebrow. But what about women in their fifties? Or same-sex couples? Or women who elect to freeze their eggs in their thirties in the hope they might be able to use them in their forties?

The ways in which the procreative desires of couples and individuals have interacted with these larger social and political forces as these new technologies developed are the focus of this book. We examine the unprecedented means—liberating for some and deeply unsettling for others—by which families have come to be formed with the advent of assisted reproductive technology. To do so, the book traces three interconnected narratives that have shaped the history of assisted reproduction in the United States. The first examines the development and expansion of these technologies from the perspective of researchers and practitioners. We begin with the work of John Rock, who with his research assistant Miriam Menkin reported the first ever fertilization of a human egg "in glass." We consider the experiences of the men and women who were among the first wave of researchers and prac- titioners in the field—some of them veteran reproductive endocrinologists, others young, untried, and eager—situating them in their larger cultural and political worlds. We also examine the structural development of IVF in the 1980s, when academic medical centers led in setting expectations for research and practice, and we consider the varied experiences of the physicians who developed programs across the United States.

The second narrative focuses on the women and couples who have used these technologies and the circumstances that affect their procreative deci- sions. Studies over the past decade have shown that between ten and twelve million Americans of conventional reproductive age today—about 12 per- cent of women and about 7 percent of men—visited a physician because of infertility at some time during their reproductive years.[9] Those who want children and cannot have them may suffer profound and enduring emotional and mental suffering. For many people, having their own biological children is critical to their beliefs about the intrinsic meaning of family, their desire to pass on that family's love and legacy, their sense of their place in the commu- nity, and their purpose in life. And these parental longings are not limited to infertile heterosexual couples. Today, the range of patients being treated in the nation's fertility centers includes, in addition to married or partnered heterosexual couples, same-sex and transgendered couples, single men and women, and women beyond their natural reproductive years.

Both stories—of scientists and physicians, and of hopeful or frustrated would-be parents—intertwine with the third, which weaves together the cultural, economic, ethical, and political forces that form the larger environ- ment within which these reproductive technologies are developed and used. The history of assisted reproduction in the United States is one of contro- versy and contention, perhaps because it raises questions that reach to the core of what it means to have—or to be—a family. Every other developed nation has addressed the medical and ethical issues surrounding these advanced reproductive technologies by creating national policies and regulations and by periodically revisiting those policies. The United States has not. Around the world, this country is known as the "Wild West" of reproductive medicine.[10] How did this happen?

We trace the uniquely American response to the development of the new reproductive technologies back to the 1970s and the charged political climate that resulted from the US Supreme Court's decision to legalize abortion in 1973. Because of the intensity of the condemnation of the court's ruling by abortion opponents, the federal government was reluctant to take a position—pro or con—on the question of allowing such agencies as the National Institutes of Health (NIH) to fund any studies involving human embryos.[11] To the organized anti-abortion movement, destroying an embryo was equivalent to murder. In 1975, pressed by scientists in the field to fund research on human in vitro fertilization, which would necessarily include the destruction of at least some embryos, the federal government instead issued a "temporary" moratorium on such funding to allow for time to study the issue. An Ethics Advisory Board was created, but when it made the decision to support federal funding of such research, its advice was rejected. Subsequent recommendations of advisory boards and the conclusions of congressional hearings likewise fell on deaf ears. More than four decades after the temporary moratorium was put in place, the funding ban still exists, now enshrined in annual congressional appropriations bills. From that day to this, the federal government has not funded, regulated, prohibited, or approved such research. This means, for example, that the National Institutes of Health remains prohibited from approving grants for any projects in which an embryo would be destroyed. Privately funded research is conducted, but this is not the same thing as NIH-funded research, which is independent and peer reviewed and would allow for impartial and systematic assessment of existing and new technologies.[12] The inability of the United States to develop a coherent national policy on these new reproductive technologies ultimately led to a research and clinical free-for-all, and in the ensuing decades, increased political polarization has only made many reproductive specialists more wary of any federal regulation of their field of research and practice.

In terms of developing policies on assisted reproduction, what happened in the United States is the opposite of what occurred in Great Britain. There, the

1978 birth of Louise Brown, the world's first IVF baby, sparked a lively debate about how best to regulate the new technologies. That debate was followed by political action. The British Parliament created a high-level commission, which released a report in 1984 that guided the creation of the Human Fertilisation and Embryology Authority, an agency that to this day directs and regulates Britain's policies and laws related to the assisted reproductive technologies.[13]

In the United States, by contrast, every attempt at creating a national policy on reproductive technology has fallen short. Consensus continues to elude Americans as the arguments of the 1970s are recast for the twenty-first century. In that earlier era, conservative ethicists linked IVF to abortion, contraception, and sexual experimentation. Today, objection to such technologies is associated with an entire constellation of changes in reproductive behavior and attitudes. There are anti-abortion conservatives who believe that assisted reproduction demonstrates a lack of respect for what they consider life. And there are concerns among those who advocate for the rights of the dispossessed that the ways in which some of these technologies are used are inconsonant with reproductive justice.[14]

Because as a nation we have been unable to develop a consensus on the development and use of the assisted reproductive technologies, the United States by default has agreed to let the market determine what kinds of technologies are developed, how they are used, and who can access them. Are there some technologies and practices that are exploitative? Some that the society should not allow to develop? Some that should be encouraged? How can we eliminate the racial, ethnic, and socioeconomic disparities that make it difficult, if not impossible, for some groups to undergo treatment? Finally, at a time when many Americans do not have access to basic health care, including reproductive health care that might help them prevent future infertility, how should we be thinking about paying for the related costs—for high-risk pregnancies in women beyond menopause, for example—that are typically paid for by insurance?

In the mid-1940s, when the public first became aware of in vitro fertilization as a possible treatment for infertility, there was a widespread pronatalist consensus in the United States that was bolstered by an almost unquestioning faith in the power of science to improve the life of society and of medicine to enrich the lives of individuals. As a result, the idea of babies being conceived outside their mother's body engendered surprisingly little opposition. No one worried about cost or access—certainly not John Rock. Given the way he practiced medicine at the time, he simply assumed that if IVF ever became a reality, women without means would receive it for free or at a very low cost, and women of means would pay for it. By the time the baby boomers came of age in the 1970s, when it was clear that IVF babies were about to become a reality, this postwar consensus—pro-science, pro-medicine, and pronatalist—had already fractured. This time around, in vitro fertilization was more controversial, with vocal critics appearing from the left and right. A great divide opened between those who see assisted reproductive technology as merely an unconventional way to create a conventional family—and today, let's not forget, a conventional family might have two dads or two moms instead of one of each—and others who argue that it raises profound ethical and social issues and cannot be embraced unquestioningly.

The characteristics of families have changed since the 1970s, but the idea of family is as salient as ever. Today, young women who have not yet found the right partner are told they can freeze their eggs until true love comes along, and a profitable marketplace is shaping up to provide the service, even though there is no guarantee of success down the road.[15] And in June 2015, the US Supreme Court ruled that the Constitution guaranteed the right of marriage to same-sex couples.[16] There is an important corollary to these stories and related ones. For many people, marriage involves children. Today, step-families, families with two fathers and six children or two mothers and three children, single mothers or fathers, and unmarried as well as married heterosexual couples all make up American families. Those women who freeze their eggs are doing so in hopes of partners and children—someday. And despite the hand-wringing over the fact that the birth rate in the United States is declining, most people still want to be parents. Eighty-five percent of American women have children. Even before the Supreme Court's marriage equality decision, around one hundred thousand same-sex couples— some married, some not—were already raising children, a number that has grown since then.[17]

It is long past time for the United States to address the issues raised by assisted reproductive technology. An unregulated marketplace in reproduction does not serve this country well, for reasons we discuss throughout this book. As we suggest in the final chapter, we would like to see the creation of a distinctively American version of Great Britain's Human Fertilisation and Embryology Authority, which as a quasi-governmental agency is an "independent authority" overseeing both research and clinical treatment.[18] We know that, particularly in this current polarized political environment, getting Congress's attention will be difficult. But we can learn some lessons from earlier efforts. We may not have a crystal ball to foresee exactly what form a good regulatory system in this country would take, but if we do not begin now, we will likely be unable to tame the Wild West of reproductive medicine for another generation at least, and perhaps not ever.

1

TEST-TUBE BABIES JUST
AROUND THE CORNER

A young bride in a small Pennsylvania mining town took a leap of faith as she composed a letter to one of the nation's most prominent infertility specialists. The date was August 17, 1944. The recipient of the letter was John Rock, director of the Fertility and Endocrine Clinic at the Free Hospital for Women in Brookline, Massachusetts. The writer, Mrs. M. C., confided that when she was nineteen she had an operation—she did not say what for—and she emerged from surgery to learn that her fallopian tubes had been removed because of "adhesions." That operation had destroyed her dream of becoming a mother, she said, but now, "after reading on your work," she felt an upsurge of hope. Could Dr. Rock make it possible for her to bear children after all? "I thought perhaps you could help me," she said hopefully. "I remain waiting." In a touching postscript, she added, "Please answer."[1] Mrs. C. sent that letter after reading about the result of a study conducted by Rock and his research assistant Miriam Menkin on human in vitro fertilization. Two weeks earlier, the August 4th issue of *Science* had published their report of having successfully fertilized, in a glass dish, four eggs from three women. All the eggs had undergone cell division, marking the first stage of embryonic development. The news was electrifying. Within hours, every wire service and major newspaper in the country picked up the story.

Rock and Menkin were careful not to overstate their results. "First stages in the cleavage [that is, the division of cells in the early embryo] of the fertilized human egg have, as far as we know, never been reported," they began. "We believe we have succeeded in three experiments, which constitute the subject-matter of the present report," they concluded judiciously.[2] The press was not to be put off by the authors' caveats or formal language. Journalists knew a big story when they saw one. Converging on Rock, they wanted to know where this research was heading and when it would get there. In the heady aftermath of his and Menkin's success, which had been six years in the making, Rock told reporters that pregnancy using this technology was "not beyond the realm of imagination, and it seems to offer about the only hope for women whose tubes have been destroyed."[3] Thanks to the Associated Press's wire service, Rock's quote reached almost every corner of the United States, and reporters made the most of it. Readers across the country could be forgiven if they thought that what the press dubbed "test-tube babies" were just around the corner.

Mrs. C.'s letter was among the earliest to arrive at the Free Hospital, but she was by no means alone. Requests for help came pouring in from women whose fallopian tubes were diseased or absent. Each time a new article appeared, another batch of letters arrived. Some of the women mentioned appendectomies, and a few told of ectopic pregnancies, but other women had never learned exactly why their doctors had removed their tubes. One woman from California, then in her early thirties, wrote that when she was twenty-nine her doctor "said her tubes were dried up so he removed them" during an appendectomy. She hoped that in vitro fertilization would offer "a modern surgical miracle" that would allow her to have another child. Some women asked for particulars. Mrs. P. C. H., of Richmond, Virginia, had lost both tubes and one ovary during surgery. Her letter noted that she had "often thought of something in an experimental way," and she sent a list of questions asking exactly how this new technology might allow her to become pregnant.[4]

The fallopian tubes are fine, narrow structures, partially made of smooth muscle, less than ½ inch wide and about 4 inches long, with a fringe of tissue at the ends to capture the newly released egg. Tubal disease, from which many of these women suffered, was a common cause of infertility. Around 20 percent of Rock's infertility patients suffered from this problem; other doctors estimated the overall proportion of infertile women with tubal disease as high 47 percent. The condition was caused by endometriosis, sexually transmitted infections, or scarring arising from complications of pelvic or abdominal surgery. Tubal factor infertility was exceedingly difficult to treat, although it had become easier to diagnose using a procedure developed in the 1920s by fertility expert I. C. Rubin called tubal insufflation, in which carbon dioxide was introduced into the uterus and then blown through the fallopian tubes in order to detect obstructions. Small bits of debris in the tubes could also be pushed out by the carbon dioxide, and sometimes this was enough to allow fertilization to occur. But the only treatment for an actual blockage was an operation to try to reopen the tubes, a surgical procedure requiring extraordinary skill and delicacy.[5] Success rates for tubal repair were low. The most highly skilled surgeons, including Rock, reported pregnancy rates of only around 7 percent. If a way could be found to bypass the tubes, many of these women might be able to bear children.[6]

Having specialized in the treatment of infertility for two decades, Rock understood the anguish these women felt in dealing with a problem that could rarely be solved by the methods doctors had at their disposal, and he answered their letters with kindness and sympathy. To women in their late twenties or thirties, he offered no hope that in vitro fertilization could help them. The research simply wasn't far enough along. He suggested that they consider adoption. To one woman who confided that infertility and her husband's refusal even to discuss adoption had caused her to fall into despair, Rock sought to comfort her by suggesting she seek other ways to fulfill her life. "Child-bearing," he wrote, was not the "only purpose for which a woman is placed on this planet." But when the letters came from women in their early twenties, his initial responses conveyed a sense of optimism. In the first months after the first fertilizations, Rock dared to hope that clinical applications might follow within a decade or so. When he wrote Mrs. C. to share his deep regret that "our research work has not progressed to the point where it is of any clinical value," he could not help himself from adding, "Fortunately you are young yet so don't give up hope." In the future, he anticipated, in vitro fertilization would make it possible for women in her situation to become pregnant and bear children, and that future might come when Mrs. C. would be in her early thirties and still well within her reproductive years. Unfortunately, his prediction was off by more than three decades.[7]

An Unlikely Journey: From Banana Plantation to Medical Career

John Rock was fifty-four years old and in the prime of his career when what the press called "test-tube fertilization" catapulted him to national prominence. A story soon began to circulate that "a famous movie actress"—almost surely Merle Oberon—had arrived in Boston in the aftermath of all this publicity to consult "the fertility doctor," but she apparently didn't know his name. After asking at one of the hospitals, she was sent directly to Rock. When they met, and she told him a little shamefacedly how she found him, he told her not to feel embarrassed. After all, he said, "Before we met, I'd never heard of you, either."[8] To Rock, a movie star was just one more person in need of help. He was used to treating women from all walks of life, with a patient roster that already spanned the social hierarchy and ranged from the wives of laborers and elevator operators to Boston socialites. By all accounts, his patients found him to be a caring and attentive doctor.

Born in 1890, John Rock was the youngest son and next-to-youngest child in his family, coming into the world just minutes before his twin sister. Both his parents were the children of Irish immigrants. The family lived in Marlborough, Massachusetts, a small industrial city about 30 miles from Boston. His mother, Annie, raised the children. His father, Frank, was a liquor dealer and speculator. When the town enacted anti-liquor ordinances in 1908, Frank told his sons that they would have to support the family—Frank, Annie, and their two sisters, one of whom was in college—until he could find a new line of work and get back on his feet. The older sons were already working. John was finishing up a postgraduate year at the High School of Commerce in Boston; as soon as he graduated, Frank shipped him off to Guatemala to work for the United Fruit Company as a timekeeper on a banana plantation.

At the age of nineteen, John found himself responsible for managing the labor of two hundred or so Jamaican and indigenous workers. Taking this job had not been a choice for him, but his family needed the money. He was unhappy. He hated the heat, disliked what he considered the crude behavior of several of his fellow clerks and managers, and disapproved of the way some of them treated the laborers. That is not to say that John was progressive on racial issues or was opposed to the United States' role in empire building in Latin America. After all, he arrived in Guatemala straight from studying at an elite public school created to teach young men how to succeed in the modern world of business. If anyone had ever asked him about his views of the morality of the imperialist system of which he was about to become a part, he would have been bewildered. Like many whites, he accepted racial separation and inequality as part of the social order. But witnessing racism in person was different. John found the blatant form of racial domination that was practiced here—what one historian called United Fruit's "Jim Crow framework of 'negro management'"—appalling.[9] About the only person he considered a friend was the company's twenty-six-year-old Scots physician, Neil MacPhail. The two young men took to each other; John was soon spending most of his free time at the small hospital that the doctor managed. Because he was there so often, and the doctor was rarely off duty, John began to assist MacPhail in the operating room.[10]

John Rock's experience in Guatemala would chart his future course in life, although not in the way his father had hoped. He had been in his new job for just a few months when the company ordered a wage cut for the field workers. John thought the pay cut was wrong, so this teenaged timekeeper—perched on the lowest rung of management one could imagine—took it upon himself to refuse to follow management's orders. His quixotic defiance, of course, had no impact. The company cut the wages anyway. The result was worker outrage, a wildcat strike, and the arrival of Guatemalan troops, who were sent in to quash the strike. A tense situation ensued, made more dangerous when a white employee, with no provocation, shot three of the Jamaican workers, one critically, during the protest. A melee ensued, and one of the Americans nearby sent for John and MacPhail. Both men came quickly, waded into the angry crowd, and rescued the wounded. While the doctor took the casualties to the hospital, John stayed behind to see if he could calm the situation. He apparently succeeded; after all, the workers knew that he had taken their side in the dispute. So did the company, which soon fired him.

John's time in Guatemala affected him deeply. It was here that he learned to defy accepted behavior and reject conventional attitudes when they conflicted with his own sense of right and wrong. It was a lesson he never forgot. He learned he had the courage to stand his ground and make his own ethical judgments. He also decided to follow in the footsteps of his friend and become a doctor. Medicine, he concluded as he watched MacPhail at work, required an open mind, independent thinking, quick judgment, and confidence in one's decisions. Being a physician and being true to his own beliefs merged in his mind, and he returned home ready to pursue a medical career.[11]

His father and oldest brother, Charlie, however, were not so ready. To please them, John took a job as a clerk. His heart was clearly not in it, and he was fired again. Charlie, realizing at last that his youngest brother was surely not cut out for a career in business, agreed to send him to college, and in 1911, John entered Harvard. He finished his undergraduate requirements in three years and stayed on for medical school. After receiving his MD in 1918, he completed an internship in surgery followed by residencies in gynecology, urology and urological surgery, and obstetrics.[12] By the time his training was over, he was thirty-one years old. Later, he would joke that he had to go into obstetrics and gynecology because he was getting too old to do anything else, but in truth, he had found his calling. In 1924, he was invited to reopen a defunct sterility clinic at Massachusetts General Hospital and to help organize a new one at the Free Hospital for Women. Both hospitals were affiliated with Harvard, and John received a faculty position at the medical school. In 1926, now married and the father of the first of his five children, he was named director of the Sterility Clinic at the Free Hospital for Women. He would remain there for the next thirty years.

The Fertility Doctor

In the first two decades of the twentieth century, medical interest in what was then called sterility was limited. The surgical therapies of the late nineteenth century had proven largely ineffective, and doctors had little else to offer. By the 1920s, however, new developments in both diagnosis and treatment were starting to make their way into clinical practice. Still, there were many unknowns. A doctor who suspected that pelvic disease was causing a patient's infertility could view the pelvic organs only by cutting open her abdomen in a major surgical procedure called a laparotomy. Physicians had no means of predicting accurately when a woman would ovulate. They understood only dimly the process by which a human sperm fertilized an egg, and they knew even less about such things as the length of time it took a newly fertilized egg to find its way into the uterus and implant successfully. And although—thanks to the work of George Corner, Edgar Allen, and Edward Doisy—scientists had learned about the existence of estrogen and progesterone, the so-called female sex hormones that regulate the menstrual cycle, no one had yet figured out how to isolate the latter or how to synthesize either of them.[13]

John Rock was a clinician, not a basic scientist; his interest in these discoveries lay primarily in how they might help practitioners understand the reproductive cycle and treat disorders associated with it. With that end in mind, in the 1930s he developed an ambitious research program that began with pioneering studies on the timing of ovulation and the function of the endometrium (the lining of the uterus) during the phases of the menstrual cycle. This research, conducted with a younger colleague, Marshall Bartlett, made it possible for the first time ever to determine—although only after the fact—if and when a woman had ovulated.[14] Rock renamed the Sterility Clinic, calling it the Fertility and Endocrine Clinic, to reflect the growing body of new knowledge in the field. He also undertook two studies that would illuminate the earliest stages of fertilization and the process of implantation. The first, in collaboration with Harvard pathologist Arthur Hertig, tracked the early stages of human embryonic development. The second was the one with which we began this chapter, the research with Miriam Menkin on the fertilization of human eggs outside a woman's body. Taken together, Rock's research with Hertig and Menkin in the 1930s and 1940s established an extraordinary body of clinical research that helped to lay the foundation for the development of the field of reproductive medicine during the next half century.

Hertig and Rock's embryo study, as it was called, documented the process by which a fertilized egg makes its way into the uterus and implants. At the time, long before the development of ultrasound, the interior of the body and the processes that occurred in the ovaries, fallopian tubes, and uterus remained largely a mystery. No one knew exactly how a woman's eggs were fertilized, or exactly where, nor did they know at what stage of conception the fertilized egg implanted in the uterus, or in what part of the uterus implantation occurred. Neither was it known what proportion of fertilized eggs were abnormal and therefore destined to miscarry. This research, conducted by Hertig and Rock from 1938 into the early 1950s, provided critical baseline knowledge about the stages of early embryonic development and informed the work of researchers in reproductive medicine for decades to come. In 1989, a year before he died at the age of eighty-six, Hertig expressed gratification that this work had made a significant impact. "I have been told by experts in *in vitro* fertilization," he said in a retrospective essay about this study, "that this series of naturally occurring human ova laid the foundation for their pioneering work in solving developmental aspects of human infertility."[15] The Hertig-Rock study provided the first visual record of early human fetal development ever seen, and its images were used in obstetrics and gynecology textbooks up through the early 1980s.[16]

In the same year that Rock began the embryo study with Hertig, he hired Miriam Menkin to work with him on human in vitro fertilization. The achievement of what Rock had once called "conception in a watch glass" was more than six decades in the making. In 1878, Austrian scientist Samuel Leopold Schenk reported the first ever in vitro fertilization and cell division of the eggs of rabbits and guinea pigs.[17] Over the course of the next half century, several other scientists attempted different forms of artificial fertilization or implantation. In 1890, Walter Heape successfully transferred embryos from one breed of rabbit to another. And in the early twentieth century, some of the most widely publicized scientific stories involved Jacques Loeb's research on parthenogenesis—that is, creating embryos from eggs without using sperm to fertilize them—in the sea urchin.

Loeb, who was among a handful of early twentieth-century scientists who became public personalities, never went quite so far as to create fatherless sea urchins, although he was sure it could be done.[18] He was widely revered, and his favorable public image remained unaffected even though some of the more sensationalist newspapers suggested that he was "manufacturing life," or that he was a real-life version of "Faust or Frankenstein." But even those hyperbolic accounts expressed little anxiety about his work's potential application to humans. Loeb was profiled in *McClure's Magazine* and appeared on the cover of *Harper's Weekly*. His biographer, historian of science Philip Pauly, said that Americans at the turn of the century were unworried about the ramifications of Loeb's research because they found such scientific and technological advances enthralling rather than threatening. For the public, "reproductive manipulation" was just one of many "technological possibilities" that percolated through modern life, not an idea fraught with moral questions or problems. As the new century wore on, Loeb became even more of a celebrity, serving as the model for the protagonist of Sinclair Lewis's acclaimed 1925 novel *Arrowsmith*, a story of scientific idealism lost and recovered in the competitive world of medical research.[19]

This unquestioned veneration for scientists like Loeb did not outlast the 1920s, however. The Great Depression of the 1930s unleashed a wave of pessimism about social and economic progress. This darker view of the future also affected attitudes toward science and technology, as revealed in the fate of one brilliant young biologist who sought to follow in the footsteps of Schenk and

Loeb. Loeb was a hero to the young Gregory Pincus when he was growing up in an immigrant Jewish farming community in southern New Jersey. "Goody" Pincus, as he was known—his nickname short for his middle name, Goodwin—originally planned to become an agronomist, as had two of his uncles, but he soon become much more interested in mammalian reproduction.[20] After earning a PhD from Harvard in 1927 and completing two postdoctoral fellowships, he became an assistant professor there in 1931 and began to conduct research on both parthenogenesis and in vitro fertilization in rabbits. His original mind, dazzling research agenda, and air of self-confidence earned him admiration in some quarters but raised suspicion in others. In the mid-1930s, he reported that he had successfully produced baby rabbits using in vitro fertilization. These IVF experiments were controversial enough, but he also claimed success in parthenogenesis, which the press soon dubbed "Immaculate Conception."[21]

In the 1930s and 1940s, journalists who covered science and medicine regularly attended scientific meetings for story material. At one of these meetings in 1936, they heard Pincus give a talk on his successful in vitro fertilization in rabbits. *New York Times* reporter William Laurence picked up the story, informing readers that what could be done in rabbits would likely be done in humans. Laurence let his imagination take flight. Not only did Pincus's work raise the possibility of human babies being conceived outside the body, Laurence speculated, but it could also lead to children being "brought into the world by a host mother not related to the child."[22] Given the times, it was inevitable that the reporter would bring up Aldous Huxley's best-selling 1932 novel *Brave New World*. "'Bottle babies,'" Laurence said, "predicted . . . for the distant future in a 'brave new world,' where children will be born in test tubes, have been brought at least part way toward actuality by Dr. Gregory Pincus at the Harvard Biological Institute."[23]

This was not the kind of notice that Harvard professors sought. As his biographer Leon Speroff put it, "Pincus did not seek publicity." It "found him." It found him with a vengeance the following year, when an infamous profile of Pincus appeared in *Collier's*, a popular national magazine, complete with photographs that virtually screamed "mad scientist." The author of the article, freelance writer J. D. Ratcliff, spun a dystopian fantasy in which Pincus's work on in vitro fertilization and parthenogenesis could easily lead to a reproductive future for humans in which women ruled and men were no longer necessary.[24] The article was a disaster for the young biologist. He was finished at Harvard. He was denied tenure and lost his job. But the publicity over his "fatherless rabbits" was not the only reason for his ouster. Although it played a major role, there were other factors as well. One was the rampant anti-Semitism of the Ivy League. Another was the fact that Pincus's mentor was feuding with Harvard's new president and could not help him. Pincus was brilliant and his work was groundbreaking, but those facts didn't matter.[25] He had no other prospects and soon left for England and a sabbatical at Cambridge University.

Meanwhile, across town at the Free Hospital for Women in Brookline, John Rock was thinking about how to apply what Pincus had accomplished in rabbits to human conception. In 1937, an anonymous editorial appeared in the *New England Journal of Medicine*. "The limiting barrier between what we know and do not know—between scientific knowledge and ignorance—regarding the field of reproduction . . . may give way entirely as a result of a recent discovery," the editorialist wrote, and "the 'brave new world' of Aldous Huxley may be nearer realization." A voracious reader of popular fiction, Rock was no more able than the journalists who covered the Pincus experiments to resist invoking the imaginary future society of *Brave New World*, when nearly all humans were fertilized in test tubes and spent their prenatal lives in artificial wombs, fatherless and motherless. And the "recent discovery" referred to the successful birth of a rabbit after Pincus fertilized ova from one rabbit "in a watch glass" and then implanted the resultant embryo in another rabbit. If conception in a watch glass could be made to work in humans as well, the author exclaimed, "What a boon for the barren woman with closed tubes!" The ornate writing style surely gave Rock away to those who knew him well, but it wasn't until a few years later that he admitted to having been the anonymous editorialist.[26] An IVF rabbit had killed Pincus's Harvard career, but Rock was unworried about his own.

Having credited the Pincus experiments as his inspiration, Rock then hired Miriam Menkin, the younger man's former research technician, to help him. A talented researcher, Menkin had hoped to earn a PhD in biology, but between financially supporting her husband through medical school and residency, and then having two children, she was never able to find the time to pursue that advanced degree. Instead, she worked in jobs with such titles as research secretary and technician, helping others to achieve their research goals rather than setting her own. Menkin had worked for Pincus on his rabbit IVF studies. When he lost his job, she no longer had hers. Called in for an interview with Rock in 1938, she remembered how much she wanted that job and how disappointed she was during their talk that Rock did not seem impressed by her credentials. Exasperated, she finally told him that she had prepared "the pituitary extracts FSH and LH, for Dr. Pincus's experiments on rabbit eggs." For good measure, she said, she told him that she was experienced in "SUPEROVULATION! [T]hat one word did it. . . . I was hired." She started work almost immediately.[27]

Menkin was smart, tenacious, and meticulous—a perfect fit for her new boss. Rock was brilliant, intuitive, and driven to seek solutions for the medical problems that beset his patients, but he had little patience for the tedium of the laboratory. In contrast, the laboratory was Menkin's natural home. She was committed to basic research and soon became a true collaborator on this project. And even if, like so many women in science in this period, she never had a title that acknowledged her skills, she did at least work for someone who made every effort to ensure that she'd have professional recognition. Once this project ended in success, Rock insisted that Menkin be named lead author for the detailed article on the research that would appear in 1948, hoping that she would be given proper credit her extraordinary work. He may have also lobbied, unsuccessfully, for Harvard to allow her to use this research as a dissertation so that she could earn her PhD.[28]

Rock respected Menkin's abilities and appreciated her quirky sense of humor. Menkin liked to tell people that she was Rock's "egg chaser." That description wasn't far off. The operating room at the Free Hospital was in the basement, so on Tuesdays, when one of the study volunteers was having surgery, Menkin would stand just outside the operating room. When the operation was over, if it had yielded any ovarian tissue, she took it and immediately ran—literally ran—up the three flights of stairs to the lab to look for eggs. Except for a brief maternity leave to have her second child, she did this every week. Only rarely were there any eggs to be found, although the study was never short of volunteers. Nearly a thousand women agreed to be a part of the study over the course of six years, during which time Menkin found eggs in the ovarian tissue of just forty-seven of them. And even when she did find the eggs, that was just the first step; she then had to try to fertilize them using sperm "left over," as she put it, from husbands in couples using artificial insemination.[29]

For six years, she experienced failure after failure. But then, on February 6, 1944, Menkin fertilized her first egg. Fatigue, she recalled later, was the major reason for her success. Awake for two straight nights with her teething baby, she had been too tired to perform some of the more time-consuming elements of Rock's protocol. She usually washed the sperm three times, she later recalled, but that day she did it just once because she didn't feel up to the more tedious task. Then she inadvertently used a more concentrated sperm suspension than usual. Finally, she extended the contact time between the sperm and egg. She hadn't meant to do so, but "I was so exhausted," she remembered, "that I couldn't get up." So she sat for at least an hour, alternately dozing and watching. Eventually, she went to the bench one more time. Looking into the microscope, she thought she was seeing double, so she called one of the doctors in to look. She was not seeing double: the egg had divided into two cells, an early sign of fertilization.[30]

Overjoyed, she called out to the other offices and labs, and everyone came to look at the newly fertilized egg. The excitement over her success quickly led to a group argument about the best way to preserve the specimen. They decided to use the procedure developed by Chester Heuser of the Carnegie Institution. However, in the heat of the discussion, Menkin forgot to have the egg photographed, although she had remembered to make a sketch. When the argument was over and a decision reached, she went back to the microscope to begin the preservation process, but she couldn't find the egg. She was crushed, but Rock, who was famously unflappable, consoled her. Don't worry, he said, at least now we know it can be done. Menkin went back to her weekly vigils outside the operating room; because it was unclear which of the accidental new factors had made the difference, she and Rock agreed they all would be part of the protocol from now on. She soon achieved success again, this time with three ova from two women.[31] Each of these fertilized eggs was carefully photographed before being preserved.

Rock now felt confident enough to report the results to *Science*.[32] The fertilized eggs (early-stage embryos) were taken to the Department of Embryology at the Carnegie Institution of Washington, located in Baltimore. There, several of the leading experts in mammalian reproduction, including George Streeter, the greatest embryologist of his generation, examined them. "Dr. Corner, Heuser, and I," Streeter told Rock, "are convinced that it is the real thing."[33] Like Streeter, George Corner was one of the towering reproductive researchers of the twentieth century, and Chester Heuser, an expert on primate embryology, was widely known as a technical wizard. Their opinion seemed to settle the matter, but a decade later, the prominent zoologist Carl

Hartman told Rock, "I don't believe you ever got in vitro fertilization." It was 1954, and he wanted Rock to "go back to the problem and clean it up."[34] Such skepticism should not be surprising. Some of the scientists who questioned Rock and Menkin's results continued to be distrustful of every report of human fertilization in vitro until an actual baby was born in 1978. In the late 1940s, though, there was little doubt in the minds of scientists, journalists, and the public that Rock and Menkin had achieved the first ever human in vitro fertilization.

Answering the Hopes of Childless Couples

Journalist Joan Younger, who reported on Rock and the IVF experiments in 1945, made it a point to pay attention to the women who volunteered for the study, noting approvingly that when Rock "began asking some of the women who came to the hospital if they would co-operate" in this project, many readily agreed. Younger praised them for their altruism and, indeed, some of the infertility patients thought that even if the study would not help them, it might spare the next generation of women from similar suffering. Other volunteers had come from the Rhythm Clinic, a free birth control clinic that, because the state of Massachusetts prohibited the elective use of birth control devices, was allowed only to teach its patients the rhythm method, or abstinence during a woman's fertile period. Many of these women, who had children all too easily, were sympathetic to the infertility patients they had met waiting for their appointments at the other end of the corridor. Still others might have agreed to participate just because they were asked. As Younger wrote, "Without these women, the knowledge that human ova . . . can be handled outside the body might have been delayed another decade, another generation."[35]

Rock and Menkin's announcement of a human embryo fertilized in vitro was widely, if not quite universally, celebrated.[36] There were still a few reporters unable to forget the plot of *Brave New World* who wondered whether Rock's ultimate goal was artificial gestation, meaning fetal development in an artificial womb. No, Rock said, such technology "probably never will be developed outside the imagination of fiction writers."[37] But what about the future of what he and Menkin had achieved? Rock acknowledged, as science writer Robert Bird put it, that "the ultimate step of reintroducing the fertilized cells into the human body for completion of pregnancy . . . was a present objective of science."[38] Joan Younger also told her readers, "Don't do any thinking right now about growing babies in test tubes." After all, she reminded them, "scientists scoffed" at Huxley when *Brave New World* appeared, and "they are still scoffing. . . . [The] whole idea is fantastic, and scientists have more to do now than to allow their imaginations to roam at large through freakish fields." But no one, she concluded, was scoffing at the idea of creating a pregnancy using in vitro fertilization. "If the many-celled stage of life can be reached," she wrote, "there may be hope that test-tube fertilization will answer the hopes of many childless couples."[39]

A decade earlier, Gregory Pincus and his fatherless rabbits had been considered predictors of a future where children came into the world parentless and men could become obsolete. Now journalists were presenting the technology matter-of-factly as a means to help women with fallopian tube disease bear children. The purpose of in vitro fertilization, in this new interpretation, was simply to help to solve the "problem" of childlessness. The *Brave New World* anxieties of the 1930s apparently had evaporated. As the nation shifted its focus from the social and economic inequalities laid bare by the Great Depression to the national unity required for an all-out push to victory in a world war, a wartime economic boom reached down into the ranks of skilled workers, giving them more confidence in the possibility of continued upward mobility and financial stability. Such confidence bred optimism in other areas. The war years inaugurated an era that medical historians have called the age of "triumphal medicine." It began with the development of new antibiotics and steroids, "miracle drugs" that cured bacterial infections and made it possible for people crippled by rheumatoid arthritis to walk. If these things could be done, what else was possible? Few Americans doubted that technological advances—including fertilizing human eggs outside a woman's body, which had seemed contrary to nature only a decade earlier—would be used to promote the public good.

The war years also witnessed a mini-boom in births, anticipating the postwar baby boom, as the number of births among women over thirty-five climbed. Reassured by what seemed to be a returning prosperity, these older women were becoming pregnant in increasing numbers, bearing the additional children they had postponed having during the Depression.[40] And then, in 1945, younger men began returning from the war. Like their wives and sweethearts, as Younger observed, they were "in a hurry to begin the families so long delayed."[41] Even more importantly, as the fighting wound down, and victory seemed within reach, there was growing social pressure to make sure that the women now working in war industries would be willing to turn their attention to domestic matters once peace had been restored. Babies were essential to the new domesticity, and reproductive technology fit right into its narrative. If a pregnancy did not occur as expected, medical science, the media reported, would come to the rescue. As a result, in vitro fertilization was presented not as a threatening or somehow alien technology but as a possible way to allow women suffering from a common cause of infertility to bear children. Such pregnancies were not even possible yet, and already the idea was becoming normalized.

Finally, John Rock himself was likely a factor in the positive reception of IVF. He looked like a doctor right out of Hollywood central casting—handsome and photogenic, with a resonant voice and reassuring manner. His patients' needs and longings drove his research agenda, and he made that clear to them. In turn, they trusted him. The way Rock practiced medicine does not fit the common historical understanding of the behavior of the medical profession in this era. Scholars have looked at the twentieth century as a time when patients were defined by their diseases and physicians treated conditions, not people. Gynecology, with its even longer history of reducing women patients to objects, has been especially suspect. Its history is replete with examples of powerful men using the authority of science to establish an interpretation of reproductive health and illness that privileged medical authority and ignored women's own understanding of their bodies.[42] Rock was cast from another mold, however, and while it would be a shame if he were unique, among male physicians, he may have been one of a rare breed.

Rock was especially careful to respect the vulnerabilities of infertility patients in research studies. Yes, he said, they were conscientious research subjects because they wanted so much to conceive. Because of that, he insisted, it was essential to make sure no one was taking advantage of them. Any researcher asking "people who want babies and can't have them" to participate in a research study was obligated to ensure that "they're not being exploited—that everything that is being done is done in their interests and not just for an abstract research thing." Rock refused to compromise on this point.[43] Records from Rock's practice during this period show that his patients felt cared for and valued. It was not just Rock's own manner—although patients' letters to him show how important that was—but also the attitude of his staff that created this atmosphere. In one sad situation, for example, they arranged child care and other help for a clinic patient who urgently needed surgery but mistrusted her neglectful husband and was worried that her children would be taken away from her.[44]

This attitude may help explain why, at a time when physicians routinely experimented on human tissue and what was sometimes viewed as "surgical waste" without even thinking to ask for permission, Rock sought volunteers and provided them with clear explanations of what he intended to do. Many years later, Luigi Mastroianni, who had been a fellow under Rock at the Free Hospital in the 1950s, recalled that Rock was careful to be sure that any patient who was asked to participate in a research study understood not only what she was being asked to do but also why. "The way this man communicated with patients was something I'll never forget," Mastroianni said almost half a century after the fact.[45] Rock's patients felt valued. As journalist Joan Younger quoted one of them, "Gee, I don't see how you make me so important . . . Sometimes I think it's almost as good as being a doctor myself."[46]

Just Another Way to Create a Family

For all these reasons—the replacement of Depression-era austerity with wartime prosperity in much of the working and middle classes, the positive views of science and medicine, the beginnings of the baby boom, social pressure on women to make marriage and domesticity their highest priority, and perhaps the fact that the leader of the study was both one of the nation's leading fertility specialists and a physician known among his patients for his caring and compassionate manner—at the end of World War II, in vitro fertilization no longer evoked ideas of cold and soulless reproductive practices leading to fatherless and motherless offspring, but rather a happy nuclear family made possible by the miracle of modern medicine.

Conscientious reporters such as Joan Younger and Robert Bird made sure to include Rock's cautions and caveats about the length of time it might take for researchers to move from the first stages of fertilization that he and Menkin had achieved to the implantation of an embryo into the uterus and ultimately the birth of a baby. To Rock's distress, however, less careful observers did not hesitate to suggest that clinical applications were imminent. Over the next few years, whenever accounts of Rock and Menkin's IVF research appeared in such popular magazines as *Look* and *Coronet*, Rock would receive a new batch of letters from women asking whether this discovery could help them to become pregnant. As the letters quoted at the beginning of this chapter show, none of the women appeared to find the idea either strange or repugnant. Most of them wanted to try in vitro fertilization, and they wanted to try it right away.

Rock waged a valiant effort to explain to journalists, the public, and importunate patients that such pregnancies were not on the immediate horizon. When Mrs. T. R. M. wrote to say that she hoped that IVF would allow her to have a baby after she had experienced two ectopic pregnancies and therefore lost both of her fallopian tubes, Rock was "sorry to say," he told her, that "fertilization outside the human body . . . has not arrived at any stage in which it can offer any clinical help yet." It was "his ardent hope," he went on, "that it will [be available] sometime, in the interests of people just like you, but there is still a tremendous amount of work to be done."[47]

Scientific and technical obstacles, as well as questions about the health of any woman pregnant as a result of in vitro fertilization and the babies who might be born from it, remained to be overcome. These were questions Gregory Pincus never had to face. After all, no one expected him to worry in advance about the life or health of the rabbit into whom he implanted a fertilized egg. Nor would he have faced any consequences over the condition of her offspring, even though the "normal, healthy bunnies" one of those rabbits produced were surely gratifying.[48] In humans, however, the health of mother and baby would be of primary importance. Therefore, as Rock answered the many letters he and Menkin received over the following five or six years, his replies reflected his belief that his earlier optimism about the length of time it would take to make in vitro fertilization a clinical reality had been misplaced. In 1945, he wrote to Mrs. N. W., who had written him from the Florida panhandle, "Don't get discouraged if you are a 'very young woman.' Science moves on and the time may easily come, and perhaps sooner than you expect, when something can be done for you."[49] But by 1948, he was more pessimistic. "At age twenty-four, he wrote Mrs. J. P., "there is no need to accept absolute defeat, . . . but at the same time, you want to be very practical." Although in vitro fertilization might become a reality within a decade, he said, "there is not sufficient certainty."[50]

He also continued to try to contain the inevitable journalistic hyperbole, often to little avail. He told J. D. Ratcliff, the author of the sensational 1937 profile of Gregory Pincus who was now writing about in vitro fertilization for *Look* magazine, that its clinical use was not just around the corner: "We don't know the requirements made by the fertilized egg of its environment," he explained, "and there are a great many more things we do not know." Unsatisfied, Ratcliff continued to press until he finally got Rock to admit that "theoretically, at least[,] there are no insoluble problems." Emboldened, Ratcliff proceeded to provide his own fantasy scenario of how an IVF pregnancy might unfold: "One idea is to remove eggs from the ovaries of women with a culdoscope[,]. . . . a slender, pencil-sized tube equipped with mirrors and a light. . . . An egg taken from a woman's ovary with this instrument might be fertilized and incubated outside the body, then implanted in the same woman's womb." And for those women without a uterus, "motherhood would still be possible with egg-transfer breeding"—what we now call gestational surrogacy. After all, women sell their milk, Ratcliff told *Look*'s readers, so why shouldn't they "offer their bodies as incubators for the babies of women who are denied motherhood"?[51]

Ratcliff's article, which appeared in 1950, brought Rock a new round of sad and hopeful letters. Furious over the overblown rhetoric of Ratcliff and others in the media, Rock tried even harder to explain to these women that the technology would not be developed in time to help them. "I regret very much," he wrote in 1950 to Mrs. E. M., "that work with human eggs has not progressed anywhere near the stage to which . . . it would be of any help to you. It is not too difficult to fertilize the eggs, but . . . to keep them in normal condition until the womb is ready to take them is a major problem on which we must work long and hard."[52] By 1951, he was offering no hope at all, telling one woman, "I regret to say that the matter about which you write is still in the very theoretical stage and has nowhere near approached practicality, nor will it within, I believe, many years."[53]

Rock knew by then that it was likely to be decades instead of years before IVF could be used to achieve a pregnancy. In vitro fertilization was going nowhere, he decided. At least it was going nowhere under his oversight, so he began to look for other ways to treat his patients' infertility. He returned to the search for surgical solutions to the problem of fallopian tube disease, working with a younger colleague at the Free Hospital, William Mulligan. He sought to develop more successful ways of opening blocked fallopian tubes and keeping them open with the insertion of tiny hoods. Rock consulted with a plastics company on the possibility of developing artificial fallopian tubes, but he soon abandoned the idea as unrealistic. He became interested in developing new hormonal treatments for ovulatory disorders, another leading cause of infertility among his patients. Perhaps somewhat ironically, his hormonal regimen for the treatment of infertility ultimately led to his now-famous collaboration with Gregory Pincus that resulted in the oral contraceptive.[54]

Rock had always been a practitioner first and a researcher second. As he once told a colleague, "I am not a research man nor have I a research mind. All I know are some of the problems I want answered." He was not being modest; he was simply describing his priorities. He wanted to find solutions to the reproductive problems faced by his patients and others like them. When he found that he was unable to make further progress in developing in vitro fertilization, he moved on. Menkin was deeply disappointed. Her work on IVF had been the high point in her career, and she found his decision difficult to accept at face value. She preferred to blame interference by the Catholic Church and Harvard.[55] But we have found no evidence beyond her own belief to support such an interpretation. The fact was, Rock had been unable to advance the research beyond fertilization and early cell division. He saw no clear path forward for IVF. And since women were still suffering as a result of infertility, he wanted to move on to something he could do to help them.

No figure of Rock's stature emerged to take his place in this research, and few new developments appeared in medical and scientific journals over the next decade or so. When Landrum Shettles, then a young medical researcher at Columbia Presbyterian Hospital in New York, published a report of successful fertilization of a human ovum to the morula stage (approximately sixteen cells), journalists paid little attention.[56] In 1961, there was a brief burst of media interest when Italian researcher Daniele Petrucci announced that he had successfully cultured an embryo for twenty-nine days and that his laboratory had successfully fertilized a total of forty eggs in vitro.[57] But when the Vatican's denunciation of in vitro fertilization as unethical silenced Petrucci, the story receded. Whether Petrucci stopped conducting such research or only stopped talking about it remains unclear, however. There were rumors in 1964 that he had overseen several births resulting from IVF, but there was never a confirmation. Several years later, reports had him collaborating with reproductive scientists in the Soviet Union; those accounts, too, were never substantiated.[58]

Petrucci's silencing reflected a more negative attitude toward a technology hailed in the 1940s as a potential godsend to women with tubal disease. In 1966, *New York Times* reporter Jane Brody wrote a prescient article about a report that otherwise received little notice outside of the field of reproductive medicine. Recalling that it had been more than twenty years since John Rock "achieved the fertilization of a human egg by a human sperm for the first time outside a woman's body," Brody wrote that IVF research had pretty much been abandoned as, "one by one, the scientists turned away from the test tube embryo to other aspects of biology and reproduction." In vitro fertilization, she noted, had been both "one of the most exciting developments of the century" and one that, by the time she was writing, was "legally and morally . . . riddled with fear, disgust, and the specter of illegitimacy."[59]

Now, she said, things might be about to change. Describing new research by the young Cambridge embryologist Robert Edwards and veteran Johns Hopkins Medical School gynecological surgeon Howard Jones, she wrote that the two men were seeking to understand how ova developed prior to fertilization, and they expected before too long to attempt fertilization itself. Robert Edwards had been spending the summer of 1965 in the laboratory of Howard and Georgeanna Jones, where he took an important new step on the road to IVF. When Edwards returned to England, he sought a permanent physician-collaborator and soon found one in Manchester gynecologist Patrick Steptoe. The two men began to work together on IVF in 1968, and just one year later, in 1969, they announced that they had succeeded in fertilizing human eggs in vitro. This time, the public's imagination was captured as it had not been since the late 1940s. But now, the baby boom was over and cultural and political circumstances had changed dramatically. Unlike John Rock, Steptoe and Edwards entered an entirely different social and political environment, both in the United Kingdom and the United States. This time, there would be considerably more controversy surrounding what would come to be known as "the new reproductive technologies."

2

FROM FIRST DREAM
TO FIRST BABY

It was near midnight, on July 25, 1978, in a hospital in Oldham, England, just outside of Manchester, where thirty-year-old Lesley Brown was about to give birth to her first child. Her physician was performing a routine cesarean section—routine, that is, except for the hour, the secrecy, and the film crew. Film crew? They were on hand to document the entrance into the world of the first baby conceived using in vitro fertilization. As Steptoe said later, when he had completed the procedure, he "delivered the repaired womb out of the tummy in order to demonstrate to the cameras beyond all doubt that there were no fallopian tubes present." The absence of those fallopian tubes proved that she could not have conceived naturally. Patrick Steptoe, about to be famous as one-half of the team that brought about the world's first documented IVF birth and Mrs. Brown's obstetrician, knew that skepticism and even outright disbelief would greet this birth. He wanted there to be absolute certainty that Louise Joy Brown had been conceived as a result of this new technology.[1]

For months, the media had been avidly covering the story of the approaching arrival of the world's first "test-tube baby." In April, the *New York Post* had discovered Mrs. Brown's pregnancy, scooping the local press. Miffed at having been upstaged, the Oldham newspaper soon reported that the baby would be born during the summer at its very own hospital. Once the news was out, the parents-to-be found themselves besieged at every turn. Mrs. Brown fled her house, first staying with Steptoe's daughter and then with her own relatives. As the pregnancy progressed, Steptoe, increasingly nervous about the effect of all this stress on her health, decided to keep Mrs. Brown under his own watchful eye and admitted her to the hospital under an assumed name. He and the hospital staff tried to shield her from the press and to make her as comfortable and relaxed as possible in these extraordinary circumstances. Nevertheless, she and her husband had little peace. A local detective agency, hired by one of the newspapers, offered to pay staff members for information. Steptoe complained that "pressmen . . . circling the hospital" tried "to gain entrance by any means," dressing "up as boilermakers, plumbers, [and] window cleaners." He was sure that some employees were talking to the newspaper for money, and he protested, to no avail, to hospital administrators.[2]

In her thirty-fourth week, Mrs. Brown developed what Steptoe called a "mild but persistent toxeaemia of pregnancy" (preeclampsia), adding to everyone's anxiety. At one point, someone called in a bomb threat to the hospital. Rumor had it that the caller was either a reporter or a photographer planning to intercept the mother-to-be as she was evacuated. Although the ploy failed, other reporters camped outside, just one story below her second-floor room. Finally, the guard lost his temper. "You bastards," he called out, "don't you care about the baby?"[3] Apparently not as much as they cared about their circulation figures. News organizations also offered Steptoe and Edwards substantial financial inducements for the exclusive rights to their story, which the two men refused. This was a major scientific achievement, and they intended to publish their results in a scientific or medical journal. They did, however, encourage the Browns, who had little money, to sell their personal story to one of the newspapers. What the couple could receive, they advised the Browns, would provide a financial cushion for the baby.

Two Men and a Vision

Ten years earlier, Patrick Steptoe was contemplating his fifty-fifth birthday with little notion that his life was about to change dramatically. He had been an obstetrician and gynecologist in Greater Manchester for more than fifteen years when, out of the blue, he was contacted by Robert Edwards, a forty-three-year-old Cambridge University embryologist. Edwards wanted him to collaborate on the conception of the world's first IVF baby. (That acronym was not yet in use, however. Almost everyone except the medical profession, which generally used the technology's full name, in vitro fertilization, talked about the creation of "test-tube babies.") Edwards needed a physician as a partner in order to make the leap from basic science to clinical research. Steptoe was a perfect choice. The Manchester gynecologist was an expert in laparoscopic surgery—a rarity in Britain—and Edwards knew that this skill was critical for his purposes. The two men met in early 1968. They took to each other immediately, and no wonder. Both were iconoclasts, both had confidence in their abilities, and neither was cowed by authority.[4]

One of seven children—six of them boys—Patrick Steptoe was born in 1913 into a middle-class family in Witney, about 12 miles outside of Oxford.[5] His mother, Grace, was a women's rights activist with an interest in maternal and infant health. Harry, his father, was the town's registrar of births, deaths, and marriages. In his teens, Patrick was torn between a career in medicine and classical music. In the end he chose medicine, trained at St. George's Hospital in London, and qualified in 1939. When World War II broke out, he quickly volunteered, serving as a naval surgeon. Captured in 1941, he was a prisoner of war for the next two years. As a doctor, he was allowed to move around the camp while performing his medical duties. But then his captors realized that he was helping men to escape. Steptoe spent the next several months in solitary confinement before being released in a prisoner exchange in 1943.[6] After returning to civilian life at the end of the war, he was elected a fellow of the Royal College of Obstetricians and Gynecologists in 1948 and of the Royal College of Surgeons in 1950. Given his elite training, Steptoe had every reason to expect to be appointed to a consultancy in London. He wanted to serve at St. George's.[7] But whether the competition for these positions was simply too intense, or he was hampered by his somewhat prickly personality, or for some other reason, he failed to find a position in London. He had much better luck when he applied for a consultancy in Oldham. He and his family moved there in 1951.

For Steptoe, losing out on a consultancy in London and having to settle for greater Manchester was a blow to his self-esteem. Edwards called Oldham "the backwater of a once prosperous Lancashire mill town," views likely shared by many others.[8] Manchester was the third-largest metropolitan area in England, after London and Birmingham, but it was 200 miles north of London and even further removed intellectually from the nation's hub of medical research and treatment. In the end, however, his professional location served him well. In Oldham, Steptoe delivered babies, performed surgery, founded a birth control clinic, and even created a small sperm bank.[9]

More importantly for his future in IVF, he became a pioneer in the use of laparoscopy in gynecology.[10] Steptoe saw the enormous advantages to this new technique as soon as he learned about it. Developed by Raoul Palmer in France during World War II, laparoscopy required general anesthesia but did not involve opening of the abdomen with a large incision, making surgery much less invasive.

Gynecologists in France, Germany, and the United States were becoming skilled in laparoscopy in the 1960s, but in Great Britain, leaders of the medical establishment were extremely skeptical of its safety and efficacy. Steptoe disagreed, traveling to France to learn the procedures, investing in the latest equipment, and practicing endlessly on cadavers until he gained the necessary confidence. He went to all the important conferences in Europe and introduced laparoscopy into his surgical practice, using the technique to diagnose infertility, lyse adhesions, cauterize areas of endometriosis, and perform tubal ligations. Despite his success, most of his British colleagues remained dismissive, even hostile, toward the technique. Edwards recalled that "at clinical meetings in London, Steptoe was at best disbelieved, at worst ridiculed." As Martin Johnson, one of Edwards's first graduate students and now an emeritus professor of reproductive sciences at Cambridge, wrote in 2011, Steptoe was frustrated that "his progress had fallen on the largely deaf ears of the conservative gynecological hierarchy."[11] But if Steptoe's fellow gynecologists had no interest in laparoscopy, Edwards was fascinated.[12] He and Steptoe shared more than just their ideas; they were both, in different ways, outsiders. In Steptoe's case, that status was not, at least initially, by choice. Expecting to become a member of Britain's gynecological establishment, he instead found himself exiled to its periphery. It was different for Edwards. He was at Cambridge, the heart of the scientific establishment, but he was a born iconoclast who reveled in his working-class origins.

A dozen years younger than Steptoe, Edwards was born in Batley, a small mill town in West Yorkshire, in 1925. This was a time when class distinctions remained almost insurmountable.[13] His mother, Margaret, was a machinist in a local mill. His father, Samuel, worked for the railroad repairing and maintaining the tracks. Samuel spent long hours in the mile-and-a-half-long Blea Moor tunnel, on the line running from Settle, in North Yorkshire, to Carlisle, in Cumbria, a job that often took him away from home. When Margaret was offered the security of a "council house" in Gorton, a suburb of her hometown of Manchester, she jumped at the chance. She hoped to give her three sons—Sammy, Robert, and the youngest, Harry—a chance to better themselves.[14]

All three boys sat for scholarship examinations for grammar school, and all were successful. Sammy declined, preferring to go to work instead. Robert eagerly accepted. In 1937, he enrolled in Manchester Central Boy's High School, which had already produced one Nobel Laureate, James Chadwick, winner of the Nobel Prize in Physics in 1935. Robert aspired to higher education, but he remained proud of his roots. As a boy and young man, with his mother and brothers, he spent his summers in the Yorkshire Dales to be close to their father, working as a farm laborer and thinking that he might become an agricultural scientist. He left Manchester Central in 1943 and was soon drafted into the army. To his surprise, he was selected for officer training school, which he disliked intensely. He found "the alien lifestyle of the officers' mess" off-putting; the experience "reinforc[ed] his socialist ideals."[15] As Johnson wrote, Edwards was "a life-long egalitarian" who never lost his working-class sensibilities. Years later, he would serve as a Labour Party councilor and was always "willing to listen to and talk with all and sundry, regardless of class, education, status and background."[16]

After his discharge in 1948 at the age of twenty-three, Edwards matriculated at University College of North Wales, where he quickly discovered that agricultural science bored him. In later years, he insisted that it was the lackluster teaching at the university, not his lack of interest in the subject, that explained his bad grades. But whatever the reason, he was an undistinguished student, and his dream of a career in science was in jeopardy. He was, he said, "disconsolate" over his academic performance. "My grants were spent and I was in debt," he remembered. Here he was, "the clever, ambitious, scholarship boy who looked as if he had now fallen flat on his face."[17] He picked himself up long enough to apply to a postgraduate program in genetics at the University of Edinburgh.[18] To his surprise, he was accepted. He was not about to waste this unexpected second chance.

When he arrived at Edinburgh, Edwards became transfixed when he watched a film of Alan Beatty transferring a fertilized mouse embryo into another mouse via the cervix. It was his introduction to the idea of embryo transfer, and he was so fascinated that he decided to conduct his doctoral research on mouse reproduction, with Beatty as his supervisor.[19] He earned his PhD in 1955, at the age of thirty. Like his mentor Beatty, Edwards was one of the earliest developmental biologists to take a serious—prescient is not too strong a word—interest in genetics. In the mid-1950s, Johnson said, "genetic knowledge was still rudimentary and largely alien to the established reproductive and developmental biologists of the day."[20]

Edwards remained in Edinburgh for a postdoctoral position for two years after receiving his PhD. His experience here shaped his career. Over the next few years, he made a series of discoveries that created a foundation for his later achievements in human IVF. With his future wife, biologist Ruth Fowler, whom he had met in graduate school, he achieved superovulation in mice. With American researcher Alan Gates, he studied the maturation of mouse ova. This work, as his former student said years later, "firmly placed the young Edwards at the forefront of studies on the genetic manipulation of development." Although his first post-training research positions focused on other subjects, he did not abandon his interest in embryonic development. In the early 1960s, he tried, in Johnson's words, "to mimic in vitro the in-vivo [i.e., inside the body] maturation of eggs." Elated when he finally succeeded, he believed himself to be the first to have accomplished this feat. He was crushed to discover, after the fact, that Gregory Pincus and M. C. Chang had beaten him to it back in the 1930s and 1940s.[21]

If Edwards had thought to conduct his literature review before beginning those experiments, he would have known about this earlier research, but in his early years as a scientist Edwards was prone to a certain intellectual insularity. He never lost it entirely, retaining a lifelong penchant for skepticism about the results of the work of others. For some reason, perhaps his feelings of competition with the American who had already achieved results that Edwards expected to claim for himself, he was particularly hard on Pincus.[22] And yet in many ways, Edwards had much in common with the older reproductive pioneer. Like Pincus, Edwards planned a career in agronomy but quickly became captivated by mammalian embryology. Like Pincus, he was an opinionated outsider in the clubby world of academic science. Pincus had been shut out of an academic career. Edwards, because of his working-class roots, socialist politics, and seemingly idiosyncratic scientific interests, was an outcast in the Cambridge common room. In the end, about a year or so before Pincus's untimely death in 1967 at the age of sixty-four, he and Edwards reached détente, maintaining a cordial, if long-distance, relationship. Edwards later praised Pincus as "a man who moved mountains," adding the ultimate compliment when he said that Pincus "would have made a fine Yorkshireman."[23]

When Edwards began to collaborate with Steptoe in 1968, he held the title of Ford Foundation Research Fellow at Cambridge University, and he and Ruth were the parents of five daughters. Edwards was sure that he had the ability to achieve human in vitro fertilization, but he faced the major hurdle we mentioned at the beginning of this chapter: he was not a physician and therefore had no access to patients. Although a few of his medical colleagues agreed to provide him with human ovarian tissue when they could, for the most part, he said, "When I asked … [for] ovaries and explained what [I] wanted to do, they thought I was barmy."[24] Steptoe, in contrast, responded with enthusiasm. In addition to his expertise in laparoscopy, the gynecologist had a large cohort of patients, and he could self-fund the patient-centered part of their research out of revenues from procedures that were not paid for by the National Health Service.

The two men moved ahead quickly. In 1969, just a little more than a year after they had begun their collaboration, *Nature* published their first major paper, in which they reported having achieved the earliest stages of fertilization, meaning that they had demonstrated, visually, the penetration of the egg by a single sperm. For these experiments, Steptoe had recovered fifty-six ova from patients undergoing hysterectomies. All the eggs were inseminated, and thirty-four had "matured in vitro." Eleven of them showed sperm that had begun to move through the zona pellucida, which is the membrane surrounding the egg. Seven of the eggs appeared to have been fertilized, and two of those seven seemed to be normal.

Steptoe and Edwards claimed in this article that they were the first to show definitively that human eggs could be fertilized in vitro.[25] Really? As we discussed in chapter 1, John Rock and Miriam Menkin had been credited in the 1940s with that achievement, a point made in a commentary by anatomy professors W. J. Hamilton and T. W. Glenister. The two scientists commended Steptoe and Edwards for having brought "a new dimension to research into the first moments of human life" but dismissed their claim of having been the first to do so. Rock and Menkin, Hamilton and Glenister said, had "well-documented claims to successful experimental fertilization outside the body." So too did Landrum Shettles in the 1950s, they asserted. Some other scientists did not believe that Steptoe and Edwards had even achieved IVF. Victor, Baron Rothschild, a distinguished Cambridge zoologist, was among them. He "admire[d] the work as a preliminary experiment," he said, but their claim of having achieved fertilization, he was convinced, was "premature" because "neither the juxtaposition of the sperm and egg nuclei, . . . nor their 'fusion,' was observed by Edwards et al."[26]

Edwards struck back furiously, at Lord Rothschild in particular and more generally at the idea that he and Steptoe were not the first to have demonstrated in vitro fertilization. What he and his colleagues showed, he insisted, was fertilization in its earliest stages—when sperm met egg, attached to its zona pellucida, and moved through it into the perivitelline space. They had no intention, he insisted, of attempting to complete the fertilization. What Lord Rothschild viewed as a weakness in their results, Edwards argued, was actually a strength. Although Rock and Menkin's images had shown the egg dividing, Edwards did not find that work convincing because eggs, he said, might divide without fertilization. He argued that fertilization could be proved *only* if the sperm could be shown penetrating the egg, and that, he said, was what he and Steptoe had done.[27]

As this debate showed, it was becoming obvious that no claim to success would be universally accepted until it ended in an actual pregnancy and birth. The paper by Steptoe and Edwards was a major advance, however, both because of the work that had gone before and the sheer weight of their data. In 1970, they grew fertilized eggs to eight cells, then sixteen, and the following year, they grew them to the blastocyst stage, a particularly important milestone. A blastocyst is a five- to six-day-old embryo, representing a more complex stage of development, with one group of cells that will eventually form the fetus and another group that will cause the embryo to implant.[28] This second breakthrough was widely reported, but it still did not convince all of their fellow researchers. The problem was that "no one is sure what constitutes fertilization," Joan Arehart-Treichel wrote in *Science News* in early 1973. "It may be sperm penetration. It may be penetration through egg cleavage into a blastocyst."[29] In short, scientists remained unable to come to consensus about the exact criteria for determining, in animals or humans, whether, and if so, when fertilization had occurred. "Some eggs can divide, when appropriately stimulated," noted Jean Marx in a 1973 review of in vitro fertilization research in *Science*, "even though they have not been fertilized . . . or they may undergo degenerative changes that resemble those of a fertilized egg." The only way to prove IVF definitively, Marx concluded, would be "the transplantation of the resulting embryo to a foster mother and its subsequent development to a fetus. This criterion has been satisfied for the mouse and rabbit, but not for the human—although it has been tried."[30] The gauntlet was thrown. Some people would not believe that human IVF was possible until they saw the baby and were convinced that the baby they saw could not have arrived in the world in any other way than as a result of IVF. Imagine how Steptoe and Edwards must have felt in the early 1970s as they provided one demonstration of fertilization after the other, only to be told that it wasn't enough.

Their research faced logistical problems as well. Steptoe and the patients were at the hospital in Oldham. Edwards had his laboratory in Cambridge, 200 miles away. At the time, with no direct route between the two places, he was on the road between them at all times of the day and night. It was exhausting, but at least he had their indispensable technician, Jean Purdy, to help keep him awake. Purdy was to Edwards what Miriam Menkin had been to John Rock. She was an integral collaborator in his and Steptoe's research and "the most determined and loyal" of all his assistants. Purdy's contributions were essential to his and Steptoe's success, Edwards said. Without her, "none of our work would have been possible."[31] Edwards and Purdy wore out the back roads between Cambridge and Manchester for two years, all the while looking for a way to bring Steptoe closer to Cambridge. Finally, they found a suitable facility at the General Hospital at Newmarket, about 15 miles outside Cambridge. All they needed now was funding to make the move possible. With the endorsement of Cambridge's dean of medicine, in 1971 they applied to the Medical Research Council (the British analogue to the National Institutes of Health) for funding to support Steptoe's salary and research space. Given the progress of their research, Edwards expected a favorable response. Instead, the Medical Research Council rejected the proposal outright, expressing "serious doubts about ethical aspects of the proposed investigations in humans, especially those relating to the implantation in women of oocytes fertilized in vitro" and recommending that the researchers instead conduct research on nonhuman primates. To make matters worse, the rejection included sharp criticism of Steptoe's use of laparoscopy.[32]

Edwards and Steptoe fired back with a scathing letter. First, they said, the Medical Research Council was completely mistaken in its criticism of Steptoe's expertise in laparoscopy. "More than three thousand laparoscopies have been carried out in the Oldham General Hospital, with no mortality and with only occasional minor complications," they declared. And second, the idea that the rhesus monkey (macaque) would be an acceptable substitute for human beings, they insisted, was totally wrongheaded. When it comes to studying implantation, "mice are a closer parallel to man," they snapped. "It was a sharp letter," Edwards said with some understatement, "and soon we received an equally sharp letter back." Their plans for a research base near Cambridge "had fallen into a hole in the ground forever."[33]

This rejection by the Medical Research Council exemplified a larger professional antipathy in Great Britain not just to in vitro fertilization but to infertility research more generally. The gynecological community in the United Kingdom showed virtually no interest in the problem of infertility, an area of particular concern to Steptoe, and wondered why a new treatment for it was even needed. In the 1960s and early 1970s, as Martin Johnson noted, "overpopulation and family planning were seen as dominant concerns and the infertile were ignored as, at best, a tiny and irrelevant minority." The fixation on contraception during this period was not unique to England. In the United States, *Science News* reported in 1973, "most people would probably approve" of in vitro fertilization experiments designed to "come up with better birth control," but not for the purpose of alleviating infertility. The idea of "human test-tube babies," the article continued, "frightens or repels many."[34]

The Medical Research Council was highly critical of Edwards and Steptoe for granting media interviews on their research. As one of the scientific referees of their proposal put it, "Dr Edwards feels the need to publicise his work on radio and television, and in the press, so that he can change public attitudes. . . . This publicity has antagonised a large number of Dr Edwards' scientific colleagues, of whom I am one."[35] This rejection stung. Not only had the Medical Research Council questioned the ethics of their research in general and the surgical competence of Steptoe in particular, it also criticized the two researchers as grandstanders. At the most practical level, the rejection also meant that Edwards and Purdy would have no relief from their lengthy commute to Oldham.

Steptoe and Edwards continued to labor under the twin difficulties of distance from each other and attitudes ranging from dismissal to outright opposition from the nation's research establishment. Now they had another worry. Australian researchers appeared to be on the verge of overtaking them. One of them, John Leeton, recalled that "there developed a close but largely unknown rivalry between the British team, led by Bob Edwards, and the Australian team led by Carl Wood." Wood, whose interest in IVF had been sparked by Steptoe and Edwards's 1969 *Nature* article, went on to lay claim to the first—"albeit only a short 'chemical' one"—IVF pregnancy in 1973.[36] A chemical pregnancy is one in which human chorionic gonadotropin (hCG) can be detected in the blood after fertilization. This pregnancy did not progress to the point where it could be detected by ultrasound. Nevertheless, the fact of having achieved a pregnancy at all made it conceivable that the Australians would outpace Steptoe and Edwards. Both sets of researchers—the large team at two universities in Melbourne overseen by Wood and the British group consisting of Steptoe, Edwards, and Purdy—were accomplishing relatively consistent fertilization. After that first chemical pregnancy, however, the Australians failed to achieve additional ones.

The Australians and Steptoe and Edwards were the only two groups that were considered to be legitimately, with scientific care and the publication of results, making progress in IVF. There were others, however, who either made unsubstantiated reports of successful IVF experiments in the 1970s or had such reports made about them. Rumors swirled that Daniele Petrucci, the Italian researcher mentioned in chapter 1 who had been rebuked by the Vatican in the early 1960s for his IVF experiments, had taught his techniques to a group of scientists in the Soviet Union. According to an uncorroborated report, Moscow scientists trained by Petrucci had grown one fertilized embryo into a six-month-old fetus.[37] Those claims seem to have been unknown outside Russia, but the reports about two others—Douglas Bevis and Landrum Shettles—received considerable publicity.

In 1974, Bevis, a well-respected gynecologist and researcher at the University of Leeds in England, told physicians assembled in Hull for the Annual Meeting of the British Medical Association that he "knew" of three cases in which IVF had been used successfully to create pregnancies that resulted in full-term births. As the *Times* of London reported, Bevis said that "the three children so far born had all been normal," and one of them lived in Great Britain. Flustered by the media attention, Bevis equivocated when asked about who had performed the fertilization and implantation, although most of his listeners surmised that he himself had been involved. Within a day of making his statements, there were calls for him to "persuade whoever is responsible to publish, [because] if this is not forthcoming soon, ... people have every reason to be doubtful." One physician who knew Bevis well told the *Times* that "I have known about this for over a year, other than that I will say nothing."[38] When the *Daily Mail* reported that Bevis was indeed the person responsible for the fertilizations, he refused to comment. Bevis's wife was less reticent, telling the press that her husband thought he was talking only to his fellow physicians and did not expect his remarks to become public. The publicity rendered Bevis so distraught that he gave up research in this field and throughout the rest of his long career at Leeds never again spoke of this incident. When he died in 1994, the controversy rated only a brief paragraph in his obituary.[39]

It would be hard to imagine a greater contrast between the quiet Bevis and the flamboyant, eccentric Landrum Shettles, who also earned headlines in the 1970s. Twenty years earlier, Shettles had been considered a reputable researcher.[40] By the early 1970s, however, few of his colleagues took him seriously. Although he remained a member of the staff at Columbia Presbyterian Hospital in New York, he was apparently so unsuited to patient care that his job had been reduced to the equivalent of "an admitting nurse." Probably in an effort to have him resign, the hospital took his laboratory from him. It did not have the desired effect; he began to cadge space from others. With no defined responsibilities for patient care or research, he nevertheless remained omnipresent at the hospital. "Dressed in scrubs, his white coat floating out like a reverse shadow, he was seen around the hospital at the oddest hours, seemingly engaged in some interior mission," science writer Robin Marantz Henig reported. Behind his back, he was called "the Ghost of Harkness Pavilion."[41] But if Shettles appears to have lost much of his credibility among his colleagues, to the public he was something of a popular culture celebrity as the author, with journalist David Rorvik, of a best-selling advice book for would-be parents on ways to choose their baby's sex.[42]

Although Shettles seemed not to have published in this area of research for many years, he nevertheless was convinced that he could be the first in the world to achieve a birth using IVF, if only he had a patient. In 1973, he seized what was likely his last chance. Thirty-three-year-old Doris Del Zio, in her second marriage, wanted another child, but she had tubal disease. Two operations to open her fallopian tubes had failed. She was told about Shettles, who claimed to have been growing embryos to what he said was the ideal implantation stage of sixty-four cells. He agreed to attempt to fertilize her egg with her husband's sperm and if successful, to transfer the embryo into her uterus. He made this agreement without informing his department chair, Raymond Vande Wiele. When Vande Wiele discovered the vial of what Shettles insisted was Mr. and Mrs. Del Zio's embryo, he discarded it, ordered Shettles to cease conducting IVF experiments, and a few months later forced him to leave the hospital.[43]

An End to Pronatalism?

Few Americans were paying attention to the IVF experiments in England and Australia in these years. In contrast to the 1940s, when Americans seemed to view the possibility of embryos created in a petri dish as simply an unconventional way to create a conventional family, in the 1970s, the general public seemed largely indifferent to the possibility of an IVF baby. Part of the indifference can be attributed to the fact that early marriages with three, four, or five children were no longer the norm. The baby boom children married later and had fewer children, a demographic change accompanied by a rebellion against the pronatalism that had defined the post–World War II era. The new attitudes began appearing at the beginning of the 1970s. In 1970, *Look* magazine published an article titled "Motherhood—Who Needs It?" with the daring subtitle "A Provocative Report on What May Be History's Biggest Fallacy: The Motherhood Myth." It would have been almost impossible to imagine such an article appearing a decade earlier. Challenging the widely held idea that children made marriage happier, the article suggested that they worsened it, and once couples realized that fact, motherhood would no longer be "compulsory," and therefore "there will, certainly, be less of it." A few years later, survey data from the University of Michigan's Institute for Social Research suggested that having children decreased marital happiness. And when syndicated advice columnist Ann Landers asked her readers about their own experiences of parenthood, she was stunned when 70 percent of the respondents said that if they could make their choices again, they would have remained childless.[44] Anyone who read a newspaper, paged through a magazine, or watched television during this decade could easily assume that the major reproductive issues of the day revolved around single young women who were "on the Pill" and enjoying guilt-free—and consequence-free—sex, and young married couples who were using that same technology so that they could enjoy each other and postpone parenthood, perhaps indefinitely. The National Organization of Non-Parents (NON) provided an institutional base for anti-natalism. For a group that reached its peak in 1976 with just two thousand members before it faltered and ultimately collapsed in the early 1980s, NON received an extraordinary amount of publicity.[45]

These ideas were liberating for those who did not want to replicate the lives of their parents. They were devastating to the women and men who wanted to have children but could not. In 1965, approximately 11 percent of married couples experienced infertility, and even though this was the sunset of the baby boom, these women and men surely felt supported and encouraged by the larger society in their quest for parenthood.[46] Five years later, however, the tide had turned. The pain of infertility may be timeless, but the social and cultural climate in which women and men endure their suffering varies dramatically over time and place, and in the 1970s a woman expressing anguish over her or her husband's infertility was likely to be reminded that pregnancy was unattractive or the world was overpopulated anyway. One woman recalled her doctor suggesting with a laugh that she'd be better off childless. Infertile couples, not surprisingly, now saw themselves as all but invisible in the public discourse in this decade. Instead of empathy, many of them said, their struggles were met with indifference or hostility. The inability to conceive was viewed as a private sorrow to be gotten over, not a public issue—and surely not one to rise to the level of a public health problem. Infertile couples continued to seek solutions, but they did so in an unfriendly cultural climate, they believed. Even infertility specialists began to feel defensive when they called attention to the infertile, one of them saying that "couples should not be denied the right to have children simply because others have too many."[47]

We want to be careful not to overstate the depth of these negative views of parenthood. In the end, after all, most couples who married in the 1970s did eventually become parents. Although the birth rate dropped dramatically during this decade, only about 10 percent of women of childbearing age ended up having no children at all.[48] Still, attitudes toward parenthood had clearly changed since the end of the baby boom. The birth control pill made it possible for single women to avoid the accidental pregnancies that might have sent their mothers on an earlier-than-expected walk down the aisle. It had also become more acceptable for married women to choose to be childless or to express ambivalence about becoming a mother than it had been in the past (or would be in later decades). The Boston Women's Health Collective, a leading voice in the women's health movement and author of *Our Bodies, Ourselves*, one of the most influential publications of that movement, followed up with *Ourselves and Our Children*, published in 1978. Motherhood, the book's authors emphasized, should be an option, not a requirement, for all women. Whether single, married, or partnered with another woman, childbearing was for a woman to choose, or not.[49] The pronatalist consensus had consigned women to roles as wives and mothers and little else; sympathy for those facing infertility in this period should not blind us to its oppressiveness. The erosion of these attitudes during the 1970s was a positive development for society and for individual women seeking to break free from the normative expectation that they had only one role in life—to bear and rear children. Unfortunately, in some hands the attack on pronatalism veered too sharply into attacks on parenthood and parents themselves.

It is true that because the pronatalism of the baby boom era had limited women's opportunities to choose their own course in life, some feminists chose a career over motherhood. Others sought both parenthood and a profession. The activists in the women's health movement sought to place control over reproduction, including pregnancy and childbirth, into the hands of women, whatever their decisions; in doing so they saw themselves at odds with the medical profession. As historian Elizabeth Siegel Watkins put it, "Feminists protested a whole host of injustices perpetrated on women by the largely male medical establishment, including . . . [the] medicalization of too many aspects of women's lives, such as birth control, pregnancy, childbirth, and menopause."[50]

As far as we have been able to tell, however, a feminist argument against the new technologies was not fully articulated until the 1980s, and even then, there was considerable divergence in views within the movement. In the 1970s, for the most part, when it came to reproductive issues, access to birth control and abortion, as well as authority over pregnancy and childbirth, were the principal concerns of the women's health movement. *Ourselves and Our Children* made no mention of the ongoing in vitro fertilization experiments in its discussion of infertility, for example. The authors referred readers with infertility problems to RESOLVE, a national support group for those facing infertility that had been founded in 1974, and suggested they see their doctors. We found just one radical feminist who enthusiastically welcomed the new reproductive technologies—although only under certain conditions. Shulamith Firestone argued in *The Dialectic of Sex* that in vitro conception, especially if it could be accompanied by gestation in an artificial womb, could free women if it were part of "a new value system, based on the elimination of male supremacy and the family."[51] Overall, however, we found no consensus—positive or negative—among feminists or liberals more generally about the idea of conceiving a baby in a petri dish in this decade.

There was much more agreement on the right: conservatives were almost universally opposed to in vitro fertilization as a threat to the moral order. Creating embryos outside the womb, some of which would be destroyed, they said, was equivalent to abortion, and abortion was equivalent to murder. Moreover, they argued, IVF was the first step toward destroying the biological nature of the family. If allowed to become a reality, physician and ethicist Leon Kass contended, its use could not be confined to infertile married couples seeking to bear their own biological children. Egg and embryo donation would inevitably follow, he said, and so would the use of "gestational mothers." He deplored such eventualities. IVF, he insisted, was detrimental to "the virtues of family, lineage, and heterosexuality." Ultimately, he said, its use would lead to the weakening of "the taboos against adultery and even incest."[52]

In October 1971, the Kennedy Foundation hosted a medical ethics forum in Washington, DC, on the subject of in vitro fertilization. John F. Kennedy's brother-in-law Sargent Shriver, one of the organizers, had urged Robert Edwards in the strongest terms to participate. Edwards agreed and rearranged a commitment in Tokyo to be there. Although he knew that there was some opposition to IVF in the United States, especially among conservatives, he anticipated, at the very least, a respectful discussion. Instead, he found himself on the receiving end of a major dressing down from quarters both expected and unexpected. Edwards could have predicted what ethicist Leon Kass would say, since he had already expressed opposition to IVF. So had Princeton theologian Paul Ramsey, whose opinions were similar. But if Edwards knew he was going to be criticized, he had not expected to be personally condemned. Ramsey, Edwards recalled, delivered a "denunciation of our work as if from some nineteenth century pulpit," accusing Edwards and Steptoe of ignoring informed consent and trampling on "the sanctity of life" by carrying out "immoral experiments on the unborn." In vitro fertilization, Ramsey concluded, should be "subject to absolute moral prohibition."[53]

Ramsey "abused everything I stood for," said Edwards, who was unprepared for personal invective. He was also not prepared to be attacked by James Watson, who told the group that IVF would inevitably lead to human cloning, an outcome Watson viewed with horror. Edwards had expected more scientific detachment from this icon of science, the Nobel Laureate in medicine who co-discovered the "double helix" structure of DNA, but the older man's coolness toward him at dinner the evening before the symposium had already given him pause. Watson, he said, "was less than jocular." Nevertheless, Edwards had not expected to be branded a potential murderer. He recalled being shocked when Watson told him, in front of all the assembled researchers and ethicists, "You can only go ahead with your work if you accept the necessity of infanticide, [since] there are going to be a lot of mistakes."[54]

Edwards's astonishment at such blistering attacks soon gave way to anger, and he unleashed the full force of his temper. "I was a Yorkshireman and I would be blunt as Yorkshiremen are reputed to be," he wrote, years after the incident. His attack on Ramsey's arguments, he recalled, was rewarded by "loud spontaneous applause," which heartened him considerably. And while the *Times* of London made no mention of any applause, spontaneous or otherwise, in its story of the confrontation, its reporter did note that Edwards was defended by Howard Jones, in whose lab Edwards had improved his understanding of human fertilization several years earlier.[55] At the time, however, the views of Howard Jones were not in the majority. The hostility expressed at the Kennedy forum toward research on human in vitro fertilization reflected the broad views held by most American as well as British physicians and scientists in the early 1970s. In terms of sheer weight, expert opinion trended against the development of this new reproductive technology. In 1972, the American Medical Association called for a moratorium on research in this area. Perhaps unsurprisingly, specialists in the field of reproductive medicine disagreed. The president of the American Fertility Society, Georgeanna Jones, spoke for most of her membership when she welcomed such research.[56]

Most of the participants in a symposium featured in the *Journal of Reproductive Medicine* in 1973 agreed with Georgeanna Jones, although not all of them sought an immediate rush to human experimentation. Some argued for expanded research on nonhuman primates before attempting to achieve a human pregnancy, an idea Edwards routinely ridiculed.[57] But even among reproductive specialists who urged caution as IVF research proceeded, none expressed moral or ethical objection to the technology itself. Pro-IVF physicians and scientists had their own ethical champion in Joseph Fletcher, professor of medical ethics at the University of Virginia. Fletcher was unabashedly supportive of this new reproductive technology and took direct issue with those who inveighed against the morality of a pregnancy begun in vitro.[58] Obstetricians and gynecologists also had an interest in new techniques. As the baby boom collapsed, without a corresponding decrease in the numbers of physicians who had been trained to deliver those babies, gynecologists and infertility specialists alike saw new opportunities in the treatment of infertility.[59] In the early 1970s, however, this positive view of in vitro fertilization was not shared by the profession of medicine overall.

Opposition to IVF among conservatives intensified in 1973, after the US Supreme Court decided the case of *Roe v. Wade*, ruling that women have a constitutional right to an abortion. The decision inflamed the anti-abortion movement, whose members called their position "Right to Life" because they believed that life begins when sperm fertilizes an egg. Since it was impossible to conduct IVF research without having to discard at least some fertilized eggs, for those who opposed abortion, such research was unacceptable. Altogether, except for couples suffering from infertility and the relatively small part of the medical profession that was trained to treat them, few Americans were outright champions of the development of the new technology. And with the entire anti-abortion establishment holding steady against it, those who did support IVF had considerable difficulty when they sought to change the terms of the debate. The anti-abortion crusade that developed in the wake of *Roe v. Wade* came to dominate the reactions of the nation's politicians to this technology for decades to come. In vitro fertilization, which in the early baby boom era had been overwhelmingly perceived as a technological blessing, had become a political minefield. Several states imposed bans on fetal research after *Roe v. Wade*, and in 1975, the Department of Health, Education, and Welfare suspended any funding of human IVF research until it could convene a National Ethics Advisory Board.[60] American researchers, who believed that reason would prevail once the issues were aired and explained, decided to postpone any research or clinical development in human IVF as they waited for the ethics committee to be constituted and to rule. In retrospect, as we discuss later, that turned out to be a mistake. In the short term, this decision also meant that there would be no American entry into the race to produce the first IVF baby. That honor would go, it had become increasingly clear, to England or Australia.

Bad Days in Oldham

In England, Steptoe, Edwards, and Purdy, after failing to receive funding from the Medical Research Council, resigned themselves to remaining in Oldham and moved their clinical work to a small private institution, Dr. Kershaw's Cottage Hospital, close to the Oldham General Hospital and to

Steptoe's gynecological and obstetrical practice. Steptoe's facility with the laparoscope, and his delicacy and proficiency as a surgeon, ensured his continued success in retrieving ova. Edwards and Purdy sought to develop an optimal culture medium in which to bring the sperm and egg together and to nurture the embryo afterward. They soon began to achieve consistent fertilization and were transferring what they considered to be healthy-looking blastocysts.[61] But they could not progress beyond that step. Although they believed that they had all the right elements in place, they simply could not achieve a pregnancy. The year 1972 came and went without a single conception. Even worse, the number of fertilizations had begun to decrease. Edwards could not figure out why, and in his memoir of IVF, he called this period "the bad days in Oldham." Just what, he and Purdy wondered, were they doing wrong now? Had they made some imperceptible change in their culture fluids? Was there a problem with their new batch of chemicals? As Edwards remembered it, "We were not as successful as we used to be in effecting fertilization; [and] our attempts to establish pregnancy with blastocysts which we did manage to culture failed."[62]

To make things worse, Steptoe was suffering from arthritis in his hips, which made it painful for him to stand during surgery. This was just the first of four bad years. Everyone became discouraged, and the team felt the Australians nipping at their heels, particularly after the Melbourne group achieved their chemical pregnancy. That pregnancy may have lasted for just eight days, but Steptoe and Edwards had not even reached that stage yet.[63] Then, at last, in 1975, Steptoe and Edwards finally achieved their first pregnancy. But it did not end well. The embryo was growing in one of the patient's fallopian tubes and required surgical removal. Everyone was distressed about this ectopic pregnancy, but now they knew they could achieve a pregnancy. They pressed on, continuing to use fertility drugs to bring about superovulation, resulting in multiple ripened follicles per cycle rather than the single one nature usually produces. After placement of the embryos, the patients took a variety of hormones to make implantation and pregnancy more likely.

No matter what they tried, except for two brief pregnancies, nothing worked. As a result, in 1977, they abandoned superovulation. "It was good-bye to the fertility drugs," as Edwards put it, and they committed themselves to single-egg retrieval combined with implantation at an earlier stage of embryonic development. They also tried out a newly developed Japanese product called Hi-Gonavis, which measured a hormone in the urine, called luteinizing hormone, that predicted ovulation. Now they could tell when ovulation was about to occur, which made it possible for them to determine when to schedule the egg retrieval. At long last, everything came together. Lesley Brown was the second patient on which they used this product, and she was the first to conceive. By the time she delivered baby Louise in July, three more women were pregnant. Although one would miscarry and another bear a still-born child, the third gave birth to a boy, Alastair MacDonald, in January of 1979, the world's second IVF baby.[64]

In New York, there was an ironic coda to the swirl of publicity that greeted the birth of Louise Brown. That very week, as the worldwide media was trumpeting the birth of baby Louise, a lawsuit filed in 1974 against Columbia Presbyterian Hospital and Raymond Vande Wiele came to trial. Mr. and Mrs. Del Zio had sued the doctor and the hospital on the grounds that Dr. Vande Wiele had destroyed the beaker holding what the Del Zios believed would be *their* test-tube baby. In preparing their defense, the lawyers for Columbia Presbyterian argued that Landrum Shettles was "a quack." The recent birth of Louise Brown made it impossible for the hospital to say that IVF could not possibly work. Instead, they said that Shettles used "Model T" procedures, creating a "bloody gook" that "would have been almost a guarantee of peritonitis and a danger of death." It was surely true that whatever Shettles had in his beaker would not have made Mrs. Del Zio pregnant. But Mrs. Del Zio had believed Shettles when he told her that Vande Wiele destroyed the material because "IVF was against the policy of the National Institutes of Health." The jury ruled in favor of the Del Zios, although its award of $50,000 was much less than they had asked for.[65]

Louise Brown's good health not only made the job of the lawyers for Columbia and Vande Wiele more difficult. It also knocked out one of the most important larger arguments against the use of "test-tube" fertilization. "One fear," *Newsweek* reported, "was that the fetus might have been damaged, or even altered genetically, by the dramatic procedures attending its creation." The baby's health having been dealt with, however, other objections arose. Some researchers wondered whether the claim of IVF was a hoax, a speculation easily quashed by Steptoe's foresight in having the birth filmed—with its close-up of Mrs. Brown's fallopian-tube-free pelvic cavity. That image proved that she could only have conceived with in vitro fertilization. Of course, now that IVF had been accomplished, some shrugged it off as a mere mechanical accomplishment of no particular scientific importance.[66] For all these reasons and others that had to do with a more visceral opposition to creating human life outside the body and all that entailed, this remarkable medical milestone did not earn its originators the acclaim that they might have expected and surely deserved. It was not until 2010, more than three decades after Louise Brown's birth, that Robert Edwards received the Nobel Prize for this work. Unfortunately, neither Steptoe, who died in 1988, nor Purdy, who had died in 1985 at the young age of thirty-nine, was around to share the victory, and Edwards was too ill even to attend the ceremony.

The opponents of IVF had not prevailed. All the arguments from researchers in the United States and elsewhere who believed in the 1970s that performing IVF on women was reckless and unethical did not change the trajectory of this new technology. After Louise Brown's birth, researchers and clinicians in the United States anticipated that the federal government would lift the federal funding moratorium on human embryo research. After all, it was now evident that in vitro fertilization worked, did not harm the mother, and produced a healthy baby. Surely *now* support for such research would be forthcoming from the federal government. Perhaps no one was more eager to see an end to the ban than Pierre Soupart, who had already submitted a proposal to the National Institutes of Health and was waiting for it to be approved. He and his gynecologist-collaborator James Daniell had already begun to recruit patients for their proposed clinical study, to be conducted at Vanderbilt University.[67] They had been waiting in limbo. The societal environment may have been much less supportive to infertile couples in the 1970s than it was in the 1940s and 1950s, but the feelings of the couples themselves, as they waited for the study to be approved, mirrored those of the women who had importuned John Rock some three decades earlier.

One of these women, thirty-one-year-old Dianne Grills, had suffered an ectopic pregnancy, which required the removal of one tube. The other tube was blocked, and attempts to reopen it had failed. "Every month that goes by, well, the adhesions grow back," Mrs. Grills told *People* magazine in 1978, and her husband added, "Now our only alternative is laboratory fertilization." The couple asserted that in vitro fertilization seemed no more foreign to them "than walking around with a stainless steel hip joint or a Dacron heart valve." Another of Daniell's patients was Mary Patton, a medical technician in her twenties who was married to a Nashville police sergeant. The Pattons were African American, as were, according to Daniell, "a good percentage" of the couples they were considering for this study. In Mrs. Patton's case, a ruptured ovarian cyst had caused tubal adhesions, and reparative surgery had been unsuccessful. Sergeant Patton told *People* that his wife would "see other people having their babies, and it'd make her cry, it hurt her so."[68] Public opinion surveys taken during the period suggested that African Americans overall disapproved of this new technology on religious grounds, but Mary Patton said that she did not see a conflict between faith and her desire for children. "I don't believe an egg is a human being," she said. "If I get pregnant I'll have a baby to worry about, and that's more important."[69]

In the end, none of these couples would get the opportunity to "have a baby to worry about." Soupart and Daniell had waited patiently for the ruling of the new Ethics Advisory Board, which was scheduled to take up the issue of federal funding in September of 1978. Their hopes for a favorable review of their proposed study did not seem unreasonable, but in fact that proposal would never even be considered. The birth of Louise Brown did not end the conflicts over IVF but exacerbated them. Robert Edwards may have had no idea how prescient he was when he said after her birth (channeling a famous World War II speech of Winston Churchill), "We're at the end of the beginning—not the beginning of the end."[70] In the United States, that "beginning" took place in Norfolk, Virginia, at a new and largely unknown medical school that had just happened to hire two people who were well positioned to do exactly what they pleased in the next stage of their career. And they soon decided that what they wanted, as Howard Jones said many years later, was to bring IVF "to America."

3

IVF COMES
TO AMERICA

"It would be more satisfying to me to figure out how to do one new thing in medicine than to apply existing knowledge to 1,000 people," a young battlefield surgeon wrote to his wife from the European front in January of 1945. The surgeon's wife, a gynecologist, was at home in Baltimore, juggling her medical practice and research projects with the needs of their young children and her aging parents while he was at war. The two of them longed for a future in which they would together carve out a career and make their mark in medicine. "I do not have to sell you on this. You know it," he told her. "One new thing," he told her. He knew they could do it. "Something important."[1]

The young surgeon's wife became a leading figure in the new field of reproductive endocrinology. He had trained as a general surgeon, but because he wanted to work with her, he completed a second residency in gynecology when he came home from the war. Now, he could organize his surgical practice and research to dovetail with her medical interests. For the rest of their careers, Howard Jones and Georgeanna Seegar Jones shared an office, a laboratory, even a desk. And not quite thirty-seven years after he wrote that letter, the couple accomplished that "one new thing" they had envisioned in 1945. On December 28, 1981, the Joneses became the first in the United States to achieve a birth following in vitro fertilization when twenty-eight-year-old Judith Carr, a teacher married to an engineer, gave birth to their daughter, Elizabeth Jordan Carr.

■

America Reacts to the Birth of Louise Brown

It wouldn't be quite fair to say that Americans had been caught flat-footed by Louise Brown's birth three years earlier. After all, Steptoe and Edwards had been reporting on their progress in fertilization and implantation for the last nine years. Nevertheless, it was true that as soon as Lesley Brown's pregnancy became public, a media frenzy erupted around the globe, including in the United States. And once she delivered her daughter, and Steptoe announced that two more of his patients were pregnant, no one could deny that in vitro fertilization had moved well beyond a theoretical possibility. It was now a physical reality manifested in the birth of a healthy and otherwise ordinary (if you discounted the way she came into existence) baby girl. It is no exaggeration to call her birth a worldwide sensation. In the United States, hastily called congressional hearings addressed the implications of the successful creation of "test-tube babies," during which legislators chastised the Department of Health, Education, and Welfare (HEW) for its long delay in setting up the Ethics Advisory Board.

The congressional hearings were held a week after Louise's birth by the House Health and Environment Subcommittee of the Committee on Interstate and Foreign Commerce. Democrat Paul Rogers was the chair, but it was Tim Lee Carter, a Republican from Kentucky and a physician, who effectively defined the issue. "It seems to me," he told his fellow committee members, "that it should be our duty to start with the laws now, because we may have test tube babies popping up all around, and we should start quickly to take care of this." Members from both sides of the aisle agreed that national policies on IVF were urgently needed. Why was the Department of Health, Education, and Welfare dragging its feet, the committee members demanded to know, on the creation of an Ethics Advisory Board, which had been promised three years earlier? Witnesses from HEW were put on the defensive. The agency had failed to constitute the board under the Republican administration of Gerald Ford. Now, under Democrat Jimmy Carter, it also seemed to be in no hurry. Earlier that year, HEW Secretary Joseph Califano had slowly begun to make some appointments to the board, but eight months later the roster was still incomplete, and the board had held just two brief meetings, at neither of which was the subject of research on in vitro fertilization even raised. The congressional committee was not pleased.[2]

Secretary Califano was not the first to employ delaying tactics to avoid having to deal with the issue, nor would he be the last. For those who believed that the meeting of sperm and egg marked the beginning of human life, in vitro fertilization posed a moral dilemma. This was not a one-party issue in the 1970s. A majority of Republicans opposed abortion, but so too did a significant minority of Democrats. In 1976, when the Hyde Amendment prohibiting federal funding for abortion passed the House of Representatives for the first time—it is still in effect today—103 Democrats and 98 Republicans voted for it, and 134 Democrats and 21 Republicans voted against. Forty-four percent of the Democratic members of the House, in other words, voted with the Republicans.[3]

Ford's Democratic successor, Jimmy Carter, was personally opposed to abortion, although he accepted the Supreme Court's ruling on the matter as the law of the land. When Carter was nominated for president in 1976, the Democratic platform straddled the divide between the pro-choice majority in the party and those who remained opposed to abortion with wording that was both brief and careful. It was "undesirable," the platform committee wrote, "to attempt to amend the U.S. Constitution to overturn" *Roe v. Wade.*[4] Given these political complications, it is no surprise that Republican and Democratic administrations alike sought to evade the question of federal funding for IVF research for as long as possible.[5] After the birth of Louise Brown, however, dodging was no longer an option. Once his agency had been publicly dressed down by a bipartisan House committee, Califano knew he had to do something. As a result, a month or so after the hearings, in September 1978, he officially requested the Ethics Advisory Board to study and make recommendations regarding federal funding for in vitro fertilization and related embryo research. Many of the leaders in reproductive research and practice at the nation's academic medical centers hoped that the outcome would be favorable. Back in 1975, when the board was first promised, most medical schools had decided to wait for its recommendations before deciding whether, or how, to become involved in IVF. The result of that caution was that in August of 1978, they were overtaken by a small, little-known medical school in Norfolk, Virginia, where Georgeanna and Howard Jones were beginning a new career following their mandatory retirement from Johns Hopkins Medical School.

Eastern Virginia Medical School had been founded just five years earlier. Its guiding spirit, Mason Andrews, was an obstetrician/gynecologist and local political leader with a long history in the city. He had grown up in Norfolk, graduated from Princeton in 1940, then earned his medical degree from Johns Hopkins. After serving in the Navy during World War II, he returned to Hopkins for residency training, after which he moved back to Virginia and spent his entire career in Norfolk, where he became a prominent physician and powerful civic force. For twenty-six years, he served on the city council, two of them as mayor. He was a leading figure in Norfolk's revitalization.[6] One of his most cherished goals was the creation of this medical school, and now that it had been established, he understood that a good way to bring attention to it and the city would be to recruit a few eminent researchers. The Joneses immediately came to mind. Andrews had been acquainted with them since the 1940s, when he and Howard were residents together, and he knew that mandatory retirement loomed for the couple—she was sixty-five and he was sixty-seven. He also knew they were not ready to hang up their white coats.[7]

The Drs. Jones Go to Norfolk

Bringing the Joneses to Eastern Virginia was a bold and strategic move. Georgeanna had been a trailblazer in the field of reproductive endocrinology. Because she had the good fortune (or good sense, or both) to marry a man who recognized her talents and did everything he could to make it possible for her to use them, she was luckier than many talented women of her generation who sought to combine family and career. She and Howard, native Baltimoreans and children of local physicians, met at Johns Hopkins Medical School. And although they probably didn't realize it when they first got to know each other, they had an even earlier family connection: Howard had been brought into the world by Georgeanna's father, a local obstetrician/gynecologist.[8]

Howard, like Georgeanna, had an early introduction to the medical profession. His father was a general practitioner who often took his son and namesake on hospital rounds and house calls. He died when Howard was thirteen, and the time they spent together loomed large in the young man's memory. Medicine was almost his destiny. After graduating from Amherst in 1931, young Howard came home to Baltimore to attend medical school at Johns Hopkins. Georgeanna, who graduated from Goucher College in 1932, was a year behind him. They fell in love as students and later became engaged, but they would have a long wait for their wedding day. Johns Hopkins did not accept married residents or allow residents to marry while they were in training, so the couple did not say their vows until 1940, when Howard finished his surgical residency. Georgeanna had already completed hers. They began their careers and started a family in the shadow of World War II.

Georgeanna had her first major research success as a medical student, when she discovered that chorionic gonadotropin, then called "the pregnancy hormone," originated in the placenta and not, as was generally believed at the time, in the pituitary gland.[9] After completing her residency at Hopkins, she became a faculty member at the medical school and was appointed as the gynecologist in charge of the Gynecological Endocrine Clinic and director of reproductive physiology in 1939. She held both titles until her retirement.[10] In 1944, while Howard was at the front, Georgeanna was practicing medicine, conducting research, raising their two small children, and caring for her ill father. Howard worried from afar that her domestic responsibilities were stifling her career, and he fretted over the obligations that she had to shoulder alone while he was gone. Their future, he promised her, would be different. "It is not right," he wrote her in June of 1945, as she was coping with her father's terminal illness in addition to everything else, "that you who are so talented should have to struggle with the necessities of living for a family, but perhaps after the war this can be righted."[11]

Georgeanna enjoyed a distinguished career in reproductive endocrinology. In 1949, she discovered and described the luteal phase defect, a condition in which there is a diminished production of progesterone after ovulation that can lead to an undeveloped uterine lining, which makes it harder for an embryo to implant. This was a significant achievement. A decade later, in 1960, the American Fertility Society named her honorary vice president. The tradition honored the field's luminaries, having begun in 1955 with the naming of John Rock to the title. "Dr. Georgeanna," as she was called, was the first woman to serve as honorary vice president. In 1970, she became the society's first female president, one of the few women of her generation in the front ranks of reproductive medicine.[12] Of the two, she was the one with the stellar research reputation. Howard was a surgeon who, as might be expected of a man who was willing to make sacrifices to support his wife's career, was not afraid to defy convention in other ways.

In the 1950s, Howard worked with his colleague Lawson Wilkins, who treated babies born with congenital adrenal hyperplasia (CAH). Babies who would be considered genetically female, as determined by chromosomal analysis, could have external genitalia that looked "ambiguous, ... not clearly male or female." Wilkins discovered in 1950 that CAH could be treated with cortisone. As Howard recalled in his memoir, he "repaired [the] ambiguous external genitalia."[13] Because of his skill in this surgery, in the 1960s he began receiving requests from adults seeking what was at the time called a "sex-change" or "sex-transforming" operation. Howard and several other surgeons began performing such operations around 1965 with no fanfare or publicity. Not too long afterward, some of those who had benefitted from the surgery began to talk about it publicly. In 1966, Howard became a participant in the newly created Gender Identity Clinic at Johns Hopkins—the nation's first—to provide both surgery and ancillary services for transgender adults.[14]

By this time, Georgeanna had become a renowned reproductive endocrinologist, and if one of her patients needed surgery, Howard would often perform the operation. Neither of them, however, had displayed much interest in IVF. After all, as we noted in chapter 1, almost no one in the American medical establishment was working in this area in the early 1960s. But in 1965, Georgeanna and Howard were asked to find a place in their laboratory for a young embryologist from England named Robert Edwards. He was seeking to spend some time in a laboratory where he would have access to human ovarian tissue. Having mastered in vitro fertilization of mouse eggs, he wanted to see if he could succeed with human ova; however, because he was not a physician, he had trouble obtaining them in Cambridge. His gynecologist friend Molly Rose provided him with what he called "a small but regular supply of human ova," and occasionally other colleagues obliged him as well. But with so few eggs, he found it hard to make progress. When he talked to his wife about his frustration over this situation, she suggested he seek the advice of genetics pioneer Victor McKusick at Johns Hopkins. McKusick in turn told the Joneses that a young British embryologist with interests in fertilization and genetics wanted to come to Hopkins to study human ova. Could the Joneses accommodate him? Because Howard routinely performed ovarian resections (the surgical removal of a wedge-shaped piece of ovarian tissue) to treat women with polycystic ovarian disease, he could easily supply the eggs, and they agreed to host him. McKusick invited Edwards to spend six weeks in Georgeanna and Howard's laboratory in the summer of 1965. Edwards obtained funding from the Ford Foundation for the visit.

McKusick may have failed to tell the Joneses what exactly Edwards was going to be doing with those human eggs. Having never met them until the day he arrived in Baltimore, Edwards recalled, when he described what he wanted to do, he experienced a distinct chill from the couple. "As I outlined my ideas in more detail," he recalled, "I . . . witnessed the dubious countenance, the pursed lips." Eventually, however, he said he managed to persuade them to go along. "They rallied at last," he said, and Howard promised him the couple's full support.[15] Edwards was energized by his time working in their lab, even though, as he said disappointedly many years later, "we had failed to fertilize one human egg." On the positive side, however, he returned from this visit to their lab feeling much closer to achieving success. "I felt confident I could solve that problem," he wrote. "Why, I had only just begun." He was right: within four years, he and Patrick Steptoe reported their first successful fertilizations.[16]

In what seems to have been a bit of revisionist history, Howard Jones recalled some forty-five years later that he was convinced that he and Edwards had achieved fertilization back in 1965, although he didn't realize it at the time.[17] But Edwards himself never made that claim, even in retrospect, and he wrote a great deal in later years about his manifold attempts at fertilization. We can infer that as far as Edwards was concerned, his efforts at fertilization failed until he began to work with Patrick Steptoe. Still, as early mentors to Edwards, the Joneses were among the few American researchers who could lay claim to having helped the younger man in the early stages of his work on human IVF.

Thirteen years later, on the day before their scheduled move to Norfolk, the Joneses heard the news of the birth of Louise Brown. We can only imagine the conversation between Georgeanna and Howard as they drove from Baltimore to their new home in Norfolk. They surely recalled the time they spent with the young embryologist and considered their own part in his success. Now, for the first time in their careers, they would have the freedom to work on any area of research they chose. Mason Andrews had made that promise to them, and they knew he would keep it. Howard later remembered that even as he and Georgeanna packed up in Baltimore, they had not yet made a firm decision about their future research. But when they arrived in Norfolk on July 26, finding reporters waiting for them and clamoring to know if in vitro fertilization could succeed in the United States, Howard had the couple's answer ready. "Why not?" Howard remembered saying to them,

"All it would take would be a little money." When that statement made the news, one of Georgeanna's former patients got in touch with her to offer $5,000 in seed funding.[18]

Howard Jones told this story of the beginning of his and Georgeanna's new career for years, but it is almost impossible to imagine that the idea wasn't already in their minds. Surely, as Edwards and Steptoe came ever closer to a live birth, the Joneses must have thought about the significance of the time Edwards spent working with them in 1965. Now, with their careers at Hopkins behind them, there was little to prevent them from choosing to go in this controversial new direction if they wished. They could count on Mason Andrews to put the full weight of his enormous social capital and political skill behind them. They could count on support from the medical school itself. After all, as a new institution, Eastern Virginia Medical School did not yet have the kind of federally financed research infrastructure that was making some chairs and deans wary of ignoring the stance of the National Institutes of Health on IVF research for fear of jeopardizing their other research funding. And lastly, they could count on Edwards and Steptoe to help them.

Having announced their intentions, the Joneses set their plans in motion almost immediately. The first step was an application from the medical school to the state of Virginia for a "certificate of need," which would grant official permission for an IVF clinic at Norfolk General Hospital. As expected, furious opposition from abortion opponents erupted. Charles Dean Jr., the president of the Norfolk chapter of the Virginia Society for Human Life, promised a court fight if the permit was granted. His organization and other anti-abortion groups inundated the office of Virginia Health Commissioner James Kenley with letters and petitions containing thousands of signatures. Despite the furor, Kenley ruled in the clinic's favor, stating that in vitro fertilization would "provide another means to rectify infertility problems for those couples for whom existing solutions are not adequate or acceptable."[19]

Abortion opponents may have been the most vocal objectors to the idea of babies conceived outside a woman's body, but they were not alone. *Washington Post* columnist Richard Cohen wondered "why, in a world full of unwanted babies," we would want to make "new ones in a laboratory." Syndicated columnist Ellen Goodman was not opposed to this technology, she said, but she did admit to "qualms" about it. "Fertilization and transplant" seemed to her "no more dehumanizing than artificial insemination," she said, but "we should neither fund such a clinic . . . nor prohibit it. We should, rather, monitor it, debate it, control it. We have put researchers on notice that we no longer accept every breakthrough and every advance as an unqualified good."[20]

As abortion foes mourned the embryos that would be lost in the transfer process, and Cohen wondered why the infertile could not just adopt, Goodman was worried chiefly about the long-term repercussions of an unqualified endorsement of this new technology. Her position was echoed by the *Washington Post* editorial board. "Somewhere between 300,000 and 600,000 American women are infertile because of blocked fallopian tubes," said the editors, "and for many of them the inability to bear children is a constant personal tragedy. For these people the procedure would be a godsend. The problem with the clinic has nothing to do with them. Rather, the risks concern the road down which this procedure and the knowledge associated with it are taking society."[21]

In a last-ditch battle to stop the clinic, a conservative Virginia state legislator introduced a bill to limit IVF research. When that bill failed, the last hurdle to the clinic was cleared.[22] The anti-abortion lobby had lost the battle. It was another victory for Mason Andrews and the administrators at Eastern Virginia Medical School, all of whom had remained unruffled during the controversy. After this victory, the only real negative consequence the medical school might have faced was possible lingering public disapproval. Whatever opposition to IVF remained, it was countered by a significant demand for this service among the infertile, and it had already become clear that this was a demand one clinic alone could not begin to meet.

While the Joneses were waiting for state approval to open their clinic, they received more than twenty-five hundred requests from couples interested in the new technology. These women and men had already tried everything else in their quest for a pregnancy, and now they were eager to avail themselves of what they saw as their last chance for biological parenthood. With the way cleared and about $25,000 in private funding in hand, the Jones Clinic began taking patients in March of 1980, a little over a year and a half after their impromptu announcement to reporters. As journalist Anne Taylor Fleming reported, women seeking IVF at Eastern Virginia had to meet the following requirements: "Youth (under 35), good health, bad fallopian tubes and a husband." Fleming could have added—but didn't—money. Only those who could afford the hefty fees could take advantage of even this slim hope of biological parenthood. Medical insurers refused to cover the procedure, they said, because it was considered experimental. Even with the Joneses agreeing to donate their own time until the clinic achieved its first pregnancy, each couple accepted into the program would be charged $4,000 (about $13,000 in 2018 dollars) in other costs. Even this expense was not a deterrent. By the time summer arrived, there were five thousand inquiries.[23]

These patients were seeking access to an experimental procedure that had as yet produced only a few babies in England and Australia. As we discussed in the introduction to this book, the entire process, from egg retrieval on, was burdensome and potentially dangerous. When the Joneses began, they used natural cycles, because that is what Steptoe and Edwards had done. That may have worked in Oldham, but it wasn't working in Norfolk. As 1980 came to an end and a new year began, the Joneses decided to look for inspiration elsewhere. They knew that stimulated cycles were fast becoming the norm in Melbourne. Although the Australian group's first IVF birth in June of 1980 had resulted from single-egg retrieval, these researchers had become convinced that only with stimulated cycles, which would allow both for predictability and the availability of multiple eggs for fertilization, could IVF become a practical clinical therapy.[24] The Joneses asked the founder of the Melbourne program, Carl Wood, to share his group's protocol for stimulated cycles with them, and Wood agreed.

Georgeanna, the expert in endocrinology, reviewed Wood's advice and decided that she wanted to use Pergonal, instead of the Clomid that the Australians preferred, to stimulate the ovaries. With that sole exception, the Joneses adopted the Australian protocols.[25] They employed superovulation, meticulously monitored each patient's hormone levels, and delayed fertilization for some six to eight hours after retrieval of the egg. The new regimen worked. They announced their first pregnancy in May of 1981, and everyone was delighted when it proceeded normally. On December 28, the Joneses announced the birth of Elizabeth Jordan Carr. Mason Andrews delivered the baby by cesarean section.[26] It was an honor Andrews had surely earned. He had brought the Joneses to Norfolk and gave them carte blanche. He seems to have had no notice that the couple would decide to focus on IVF, but when they announced their intentions, he didn't blink. Deploying his enormous political clout, he made it possible for the Joneses to open their clinic and ensured that the medical school provided them with the collateral support they needed. The Joneses were grateful. Georgeanna told the *Washing-*

ton Post that she and her husband would never have had the opportunity to succeed at IVF at one of the more established medical schools. Eastern Virginia, she said, provided them with what she saw as the two essentials for this work: "freedom and institutional backing." When they announced their first pregnancy, Frederick Naftolin, the chair of the department of obstetrics and gynecology at Yale, said that what the Joneses had accomplished was "a giant, colossal, big deal."[27] Soon, his department and others would follow their lead.

Mason Andrews had been confident that hiring Georgeanna and Howard Jones would put Eastern Virginia Medical School on the academic map. He was right. Their decision to ignore the implicit federal disapproval of IVF that had been manifested in the federal funding ban turned out to be a masterstroke. When Howard and Georgeanna Jones, surrounded by unopened packing boxes in their new house near Norfolk, talked to reporters on July 26, 1978, neither they nor anyone else knew the outcome of the couple's seemingly impromptu decision. On that day, no one in the research or policy communities had any idea exactly what the federal government would do, but the Joneses, with virtually nothing to lose and everything to gain, were not about to miss their chance. Whether Americans would welcome a technology that enabled children to be conceived in a petri dish remained an open question. As one reporter put it, the public was especially uneasy about the fact that "in achieving [in vitro fertilization], doctors almost inevitably must discard some unsuitable embryos, even though they represent the beginning of life." But there were also millions of infertile couples who, journalists noted, would likely be eager to have a new option for pregnancy.[28]

The Ethics Advisory Board and the Politics of Abortion

As early as the summer of 1978, it was clear that Georgeanna and Howard Jones were going to move ahead with IVF regardless of the funding moratorium of the federal government. It was not clear whether others would follow their lead. That September, Secretary Califano had finished appointing the members of the Ethics Advisory Board and provided it with instructions. Over the next several months, the board held several public hearings and numerous private discussions. In March of 1979, it released its recommendations, agreeing unanimously that HEW should lift the funding ban and allow the NIH and other federal agencies to make decisions about financial support for human IVF research on the basis of scientific merit and research priority.[29] "Research involving human in vitro fertilization and embryo transfer," the final report read, was "acceptable from an ethical standpoint" so long as the studies were designed to "establish the safety and efficacy of [IVF] and to obtain important scientific information toward that end not reasonably obtainable by other means." The board added basic caveats about compliance with existing ethical guidelines and recommended that only "married couples" should be eligible for IVF procedures, but fundamentally, its decision gave IVF funding a green light.[30]

Equally important, in a recommendation with even broader implications the board also agreed unanimously that research on embryos up to fourteen days old should be permitted. Such research, strongly opposed by anti-abortion activists, was controversial. Nevertheless, the board believed it was justifiable. Reverend Richard A. McCormick, the sole Catholic priest on the board and a professor of Christian ethics at Georgetown University, contributed to that unanimous decision. Making a point easily recognizable to Catholic theologians, McCormick said firmly that such embryo research was "a necessary evil" to advance a greater good—the alleviation of infertility.[31]

This was a pivotal moment for the future of the new technology. If the recommendations of the Ethics Advisory Board had been accepted, they would have served as the starting point for the United States to develop national policies to govern in vitro fertilization and related technologies. Resulting guidelines would have been created to allow the National Institutes of Health and other federal agencies to fund proposals. Such research, treatment protocols, and clear benchmarks to determine progress in the field would have followed. Although Howard and Georgeanna Jones were already in the process of establishing their clinic when the recommendations were published, that clinic was not set to open for another year. If research policies had been developed, the Joneses would have followed them. So too would everyone who came after them.

Hopes were high among researchers. After the Ethics Advisory Board made its formal recommendations, just one more step was needed—approval of the board's decisions by Secretary Califano. That step was never taken. Califano had expressed reservations about IVF even before the board met, and he may never have changed his mind. In his formal charge to the Ethics Advisory Board, he asked them to consider whether IVF might "lead to selective breeding" or perhaps "to attempts to control the genetic makeup of offspring." And he worried about "the use of 'surrogate parents,' where . . . rich women might pay poor women to carry their children."[32] In addition to his own doubts about the morality of the new technology, Califano faced considerable political pressure. About a month before the Ethics Advisory Board's recommendations were made public, word had leaked out about its intentions. Antiabortion activists leapt into action, taking out an ad in the *New York Times* declaring that IVF would cause "unknowable risks to human lives" and produce "inevitable deliberate abortions." Support for such a "morally abhorrent" technology, the ad went on, would "burden our consciences, and the consciences of the tens of millions of Americans in whose name we speak."[33] The groups behind the ad also orchestrated a massive letter-writing campaign to Congress. Even though every survey taken at the time reported that a majority of Americans approved of in vitro fertilization for infertile married couples, it was the opponents of the technology who wrote to their senators and representatives. Nearly thirteen thousand letters poured into Congress; 98 percent of them opposed the Ethics Advisory Board's recommendations.[34]

Califano stonewalled. A devout Catholic, not only was he susceptible to an appeal to his religious beliefs, but he also served a president who had his own reservations about abortion. Perhaps not surprisingly, however, many years later—several lifetimes, as politics goes—Califano remembered his views differently, recalling in 2004 that his "mind was open," even "tilted toward funding in vitro fertilization research."[35] But memory is not history, and his actions in 1979 belie his later recollection. Instead of approving or rejecting the recommendations of the Ethics Advisory Board, Califano sat on the report for three months without releasing it. And when he did make the report public after that long delay, he decided to ask for an additional two months of public comment. Those delaying tactics led to permanent inaction. Before that additional two-month public comment period had concluded, the president fired Califano, not because of this issue but as part of a larger reshuffling of his cabinet. Califano's successor, Patricia Harris, simply ignored the recommendations as if they had never been made, and the board itself was soon disbanded.[36] IVF researchers, including the patient Pierre Soupart, still waiting for a funding decision before beginning his research, remained in limbo. When Soupart died prematurely in 1981, at the age of fifty-eight, he became known, sadly, as one writer said, as "the scientist who died waiting for a federal research grant that was approved but never funded."[37] In 1980, a new National Commission for the Study of Ethical Problems in

Medicine and Biomedical and Behavioral Research was appointed. Its members made a deliberate decision not even to discuss IVF; as biologist and ethicist Clifford Grobstein noted, this commission knew "a hot potato when it saw one."[38]

It had been one thing for Georgeanna and Howard Jones, who had already retired from accomplished careers, to take a risk to achieve that "one big thing" Howard had ached for when he was at the front in World War II. The same could be said for their institutional home, Eastern Virginia Medical School, which was new and little known. For the Joneses to pursue IVF had considerable upside and practically no downside. If they failed, how much was lost? They had already made their reputations, and the school would have other opportunities in the future. If they succeeded, they would cap their own careers with an achievement of major proportions, and Eastern Virginia would be on the leading edge of a revolution in reproductive technology. And that is what happened. The situation was different for many other academic medical centers. Once Califano's refusal to accept the recommendations of the Ethics Advisory Board made it clear that there would be no federal funding for work on IVF, what would researchers do? The programs in Great Britain and Australia were advancing. The Joneses had already made clear that the ban was meaningless as far as they were concerned. For many of their colleagues across the country, who had viewed the funding ban as a de facto ban on the research itself, the question now was whether they should ignore it as well. Some of them wondered if doing so might put their federal grants in jeopardy. Such anxiety was not entirely unreasonable, even though in reality the funding moratorium meant only that the federal government wouldn't pay for IVF research or clinical development. It did not make such research and development illegal.

The effect of Califano's failure to accept the recommendations of the Ethics Advisory Board was not as evident at the time as it is now. Looking back, we might be tempted to say that the American research community recognized the sharp choice that faced them—either ignore the funding ban or give up the idea of pursuing the new technologies—but in 1979, it was not yet clear how important IVF would become for reproductive medicine. Some of the most respected names in the field believed that, at best, IVF would be of limited use in the treatment of infertility. Others saw it as a nine days' wonder that would quickly fade away. Views such as these, held by some established figures in gynecology, changed over time, but such initial skepticism promoted caution. "The biggest resistance," said Alan DeCherney, one of the young IVF pioneers of the 1980s, came from the tubal surgeons. "There was a large cadre of tubal surgeons who said that IVF . . . wasn't going to work, and that it wouldn't replace surgery."[39] Martin Quigley, another one of that first cohort of young physicians who were eager to make their mark in IVF, remembered being interviewed for a faculty position by a "very prominent" department chair at a medical school in the Midwest. When he told the chair he was interested in IVF, he received a less than promising response. "Clinical endocrinology," the chair said, "that would be great, that's what we need." But "this in vitro fertilization stuff? . . . That has no part in a department of obstetrics and gynecology."[40]

IVF Comes to California

If some of their elders were cautious, however, the younger generation was eager. Some of them had been convinced of IVF's significance almost from the minute they learned that Edwards and Steptoe had achieved an ongoing pregnancy. Having seen the kind of opposition that greeted the Joneses, however, they did not broadcast their intentions. As the *New York Times* reported right after the birth of Elizabeth Jordan Carr, "Some American medical centers conducting or planning the procedure are reportedly avoiding publicity."[41] Richard Marrs, just three years out of fellowship in 1982 when he became the second to succeed after the Joneses, was one of them. A Texan who spent much of his youth on his grandfather's ranch, Marrs had learned how to make the most out of the least long before he decided to become an expert in IVF; the skills he developed on the ranch turned out to be transferrable. As he well knew, a broken-down tractor, miles from help, was not going to fix itself. If we could capture his early life in a few questions and answers, it would look something like this: Failed part on the tractor? No help for miles? Don't worry, I'll fix it myself. No money to pay for college? Don't worry, I want to be a doctor, so I'll get a job as a surgeon's assistant and pay for my education that way. (At the time, the state of Texas allowed undergraduates to work as surgical assistants.) What, you hauled me all the way out to California for a fellowship, and now you tell me you have no money to pay me? Well, I'm not leaving. I'll just moonlight to support myself until you can find the money. (The University of Southern California eventually did pay him, but not until his second year, when they made up for it by offering him an assistant professor salary.) You don't think in vitro fertilization will work and there is no money in the budget for it? Don't worry, I'll clean up this old storage closet to make a lab, and I'll ask one of the technicians to help me out, and I'll shave some savings from other grants by doing all the work myself on them as well, and I'll use the money I save for my IVF project.

Richard Marrs had shown promise as a surgeon in medical school, where he had been torn between cardiovascular surgery and obstetrics/gynecology. A rotation on the surgical team of Denton Cooley and Michael DeBakey, the latter considered by many to be "the greatest surgeon ever," had inspired him. "I loved the open heart surgery and microsurgical bypass," he remembered, but he said he hated the "mortality rate on the table, [which] was over 20 percent with open heart [surgery]. I was the lowest man on the surgical totem pole in the operating room, so I was usually sent out first [to see the family].... It was a tough thing to tell someone that your father or uncle has died on the operating table." In the end, mentored by Robert Franklin, a renowned expert on endometriosis and "one of the first tubal microsurgeons," Marrs chose obstetrics and gynecology.[42]

After completing his residency, Marrs stayed on at the University of Southern California for a fellowship in reproductive endocrinology and infertility. He was in his first year when he read about Edwards and Steptoe's first ectopic pregnancy. Marrs had never done IVF research in animals, nor, apparently, had he been a part of any of the discussions about the new technology's potential that were already starting in some medical circles. What he did know, almost immediately, was that he wanted to do IVF, even if he wasn't exactly sure how to proceed. As a fellow, he was working at the time for Oscar Kletsky, who studied pituitary tumors, and he persuaded Kletsky not just to study tumors but also to grow tumor cells in culture. He recalled reasoning that "if I can grow a tumor then maybe one of these days I can grow embryos because it's a similar environment, as I thought at the time." His foray into tumor cell culture led him to meet one of the technicians in that group, Jody Greene, who "kind of took me in." She would later help him achieve his first IVF birth. As of 2018, she was still working for him.[43]

By the time Marrs finished his fellowship in June of 1979, Steptoe and Edwards had already had more than one birth, the Australians were getting closer to their first birth, and the Joneses were planning to open their IVF clinic. He was pleased to be offered an assistant professorship at USC, but he told us that he wasn't sure he could accept it. He remembered telling the department chair, Daniel Mishell, "I would really like to but I need to be able to develop in vitro fertilization." Mishell told him that IVF was a "dead end" and "I don't want you ruining your academic career." Politely, as befitted his West Texas upbringing, Marrs replied, "Well, I understand that, but you have to understand that if I stay here I have to be able to be allowed to do that." In the end, Mishell let him work on IVF so long as it was on his own time and with his own money. " 'Just take care of your clinical responsibilities and your teaching responsibilities outside of IVF. If you have time to do that and can get some money then you can do it.' I said, 'That's a deal.' "[44]

Marrs more than kept up his end of the deal, but first, he needed to figure out how to acquire a small amount of funding. He was running "several of [Mishell's] pharmaceutical clinical protocol studies," so he asked his chair, "If I can shave any money off your budget can I put that money into my IVF fund?" Skeptical, Mishell nevertheless gave him permission, so "on his next two studies I didn't hire anybody and I did all the paperwork and secretarial work. And I shaved enough money off the two studies . . . for an incubator and a microscope." Jody Greene, the technician who had taught Marrs how to culture cells, was now working for Kletsky, who allowed her to help Marrs with the IVF research. One of the nurses, an operating room supervisor who had befriended him, helped him find a storage closet that measured about 8 feet by 10 feet, with tiled walls. It was filled with junk. He emptied it out, cleaned it up, and it became his first laboratory.[45]

Young, determined, but not really prepared, Marrs started hunting around for human eggs. Before he could hope to fertilize anything, he had to be able to find them in ovarian tissue, much as Menkin had done more than three decades earlier. He and Greene "went through the Human Use Committee and got all the approvals," he told us. "The first cases that we decided to do were women that were having sterilization procedures. I would do the tubal ligation laparoscopically and I was going to time it for ovulation to see if I could get an egg or whatever. We weren't going to get them for fertilization. I just wanted to see what the process was. So I did several cases where we timed and took the fluid into the little lab and looked under the microscope and Jody and I started figuring out how to find eggs. We saw a couple of eggs that looked good and said, 'Well, got that part down.' "[46]

Now he was almost ready for the next step, he thought, and went to hear a talk by Ian Johnston, the clinical leader of the Melbourne team that had brought about Australia's first birth in 1980. Afterward, Marrs asked to talk to him for half an hour. "Who are you?" Johnston wanted to know. "I'm nobody," replied Marrs, "but I'm doing IVF." Intrigued, Johnston invited Marrs to come to Melbourne to spend some time at his clinic. Never having taken a vacation during his time at USC, Marrs had six weeks saved up. He spent them in Australia, toggling between Johnston's program at the Royal Women's Hospital and Alan Trounson's across town at the Queen Victoria Hospital. By now, the two groups were barely on speaking terms. Trounson's group was angry at having not been included on the Johnston publication describing the country's first IVF birth.[47] The Australians may have been feuding with each other, but Marrs learned from both.

Partly out of necessity, but mostly because he believed it was important for him to understand and master the entire IVF process, Marrs said he "decided at the beginning that I needed to know how to do everything, [know] when something was going right or going wrong or whatever. And so the way I set our system up I was the only physician, and I had Jody as a lab assistant, and I couldn't pay her." One of the labor and delivery nurses also volunteered to help. She didn't get paid either. Marrs made no public announcement; he simply began accepting patients. Within months, he achieved a pregnancy and birth with his fifth couple, Korean Americans who owned a small grocery story in the Koreatown section of Los Angeles. The baby was born in June of 1982. This tiny program, consisting of one unpaid doctor, one unpaid technician, and one unpaid nurse, became the second to achieve an IVF birth in the United States. Marrs credits the Australians—Alan Trounson, Ian Johnston, Alex Lopata, and Carl Wood—for making his success possible. "There was no reason other than them that that happened."[48]

The experience of Richard Marrs was quite different from that of the Joneses and shows how IVF "came to America" by multiple pathways. The Joneses were distinguished physicians who had worked with Robert Edwards in his early years. Marrs and another physician, Martin Quigley, whose program at the University of Texas Medical Center at Houston was the third to have a birth, were young and enthusiastic. To some degree self-trained, both had honed their skills with the Australian teams. Anne Colston Wentz at Vanderbilt, whose program was the fourth to succeed, had been a fellow of Georgeanna Jones at Johns Hopkins. And Alan DeCherney at Yale, whose IVF clinic would be the fifth to have a birth, had trained under Luigi Mastroianni, John Rock's protégé. In the 1980s, all these paths would begin to merge.

In vitro fertilization came to America just as the nation was experiencing yet another shift in attitudes toward reproduction, the family, and women's roles. By the early 1980s, after a decade of celebrating the joys of a child-free life, newspapers and magazines began to feature stories chiding well-educated baby boomers for not accepting their responsibilities to reproduce. It was not yet a full-blown media panic—that didn't set in until around mid-decade— but the alarm was beginning to be sounded. The slew of warnings aimed at young women with professional ambitions represented pronatalism's coercive and anti-feminist underside. If women waited to become pregnant much beyond their mid-twenties, they were told, they just might find themselves unable to do so. These dire and overblown media predictions were not borne out by facts. Not even the statistical evidence that the infertility rate in the United States had *declined*—from 11.2 percent to 8.5 percent between 1965 and 1982—made any difference.[49] Nor did the fact that the birth rate was rising in the 1980s. The new pronatalists were not particularly interested in facts—neither the uptick in births nor the decline in infertility. The swing in public emphasis from the pleasures of childlessness to the importance of parenthood was not driven by statistics, nor was it coincidental that this shift was accompanied by a conservative political surge that was boosted by the presidential election of 1980.

The 1970s had marked the apogee of second-wave feminism. And although feminists did not speak with one voice on marriage or motherhood, they did agree on two basic tenets. One was that marriage as well as the bearing and rearing of children must always be a woman's free choice, one never socially imposed or coerced. Second was that the elimination of inequality in the workplace *and* the family would benefit all women regardless of their marital or parental situation. As the not-so-subtle attack on young professional women suggested, however, feminist ideas overall would increasingly come under attack as conservatism rebounded in the 1980s.

One early signal of decreasing support for women's rights came in 1980, when Republican Ronald Reagan handily defeated the Democratic incumbent president, Jimmy Carter. Running on what his conservative supporters called a pro-family agenda, Reagan received just over 50 percent of the popular vote to Carter's 41 percent. His electoral college victory was even more sweeping—489 votes to 49. Of course, not everyone who pulled the lever for

Reagan was a bona fide political conservative. Some voters were simply fatigued with President Carter. Others were charmed by the actor turned politician who promised them that his victory would bring a new dawn. As his ads promised the nation, once again it would be "morning in America." But if those captivated by Reagan's persona or disaffected with the status quo provided his significant margin of victory, they were not his core constituency. Reagan's bedrock support came from two different wings of the conservative movement: free-market capitalists who favored small government and deregulation, and self-styled pro-family activists who opposed abortion, feminism, and pretty much any behavior not comporting with traditional gender norms. The two groups were not natural allies, but both were willing to enter a political marriage of convenience.[50]

Pro-family conservatives, as we discuss in chapter 5, would help to drive an anti-feminist agenda that over the next decade percolated into the larger culture. For them, the word "family" had a specific meaning. For them, a family consisted of two parents, one of either sex, was middle or working class, had a stay-at-home mother and a breadwinner father, and almost always was white. They opposed federal funding of embryo research because they believed that life begins at the moment when the sperm and egg unite. The free-market conservatives who supported Reagan were less interested in the kinds of issues that animated their allies in the pro-family movement, but they were unwilling to challenge those allies on reproductive issues.[51] Moreover, their embrace of federal deregulation militated against the idea of supporting federal funding of embryo research or regulation of IVF. As a result, even though the two wings of conservatism came to their views of federal funding for the new reproductive technologies from different starting points, they ended up in the same place. Eventually, pro-family conservatives would make their peace with IVF for heterosexual couples as a treatment for infertility, but their opposition to embryo research would remain a constant.

In this context, American researchers and clinicians in reproductive medicine were beginning to understand that the opposition to federal funding of human embryo and IVF research might be long lasting. A Democratic administration had already ignored the recommendations of the Ethics Advisory Board. The Republican Party now consisted of one group of conservatives who opposed all embryo research and another who believed in the limitation of federal authority, and under the Reagan administration, the federal biomedical funding agencies became even warier of breaching the funding ban.

In a sign of the times, in 1982 an NIH embryologist was forbidden to speak—even at his own expense and on his own time—at an IVF conference on the grounds that it might violate the funding ban.[52]

By then, however, researchers and clinicians had already decided not to let the ban stand in the way of the development of the new technology. Even as the Joneses were gearing up to open this country's first IVF clinic, Marrs, Quigley, and DeCherney were readying themselves to start their own programs, which were all up and running in 1981, and programs at the University of Pennsylvania and Vanderbilt would join them by the spring of 1982.[53] It was clear to those who wanted to develop IVF programs that it would be futile to wait for the blessing of the National Institutes of Health.

Good Eggs, Bad Tubes, Youth, and a Husband

In this discussion of medicine and politics, it is important not to lose sight of the reason for the development of in vitro fertilization in the first place. Nowadays, IVF and related technologies are used for such a broad range of conditions that it is easy to forget why it was first developed, which was to treat a specific and particularly intractable cause of infertility—blocked or otherwise damaged fallopian tubes. Endometriosis, complications from abdominal and pelvic surgery, and pelvic inflammatory disease could all adversely affect the fallopian tubes, and in the United States in the 1970s there were about half a million women suffering from tubal disease. Those numbers were about to rise as a result of the use of the Dalkon Shield, a flawed intrauterine device (IUD) that had been a popular contraceptive in the early 1970s. At mid-decade, however, reports of severe pelvic infections caused the manufacturer to take if off the market in the United States. By then, more than 2.4 million women had used this IUD, many of whom had never borne children, and at least 200,000 of them contracted pelvic inflammatory disease.[54] In addition, from the mid-1960s through the 1970s, there was a rise in sexually transmitted infections among women between the ages of twenty and twenty-four, an increase reflected in this group's rising infertility rate. Even as infertility among American women overall was declining, among women in their early twenties, the rate had nearly tripled, from just under 4 percent in 1965 to 11 percent in 1982.[55]

These were the kinds of conditions, collectively known as tubal factor infertility, that in vitro fertilization was developed to treat. Young, married, and otherwise fertile women with tubal disease were the first IVF patients. England, Australia, and the United States were the first three countries to produce babies using IVF, and in each of them the first "test-tube babies" were born to such women. We know a bit less about these mothers than we do about their famous babies, which is understandable. Everyone wanted to know if the babies were healthy and "normal." Everyone wanted to see their photographs. And ever since, the media has followed the three baby girls—Louise Brown in England; Candice Reed, the first IVF baby in Australia, born in 1980; and Elizabeth Carr, born in Norfolk, Virginia, in 1981—as they grew up. Their mothers never garnered nearly as much attention, but their stories are just as important as those of their daughters. These women were the ones who had the daring to accept the risks of a completely unknown procedure and the luck, in at least two of the cases and perhaps all three of them, to become pregnant on their first attempts. Lesley Brown and Judith Carr were childless. Linda Reed had borne one baby but longed for another. At thirty, Brown was the oldest, and she had been trying to become pregnant for more than ten years. Carr was twenty-eight, and Reed was twenty-six. They were ordinary young women who just wanted to have a baby.

Lesley Brown and her husband, John, grew up in the roughest parts of Bristol in the 1950s and early 1960s. Lesley's father had abandoned the family when she was only two years old. Her mother subsequently worked low-wage jobs and had little time to spend with her little girl. As a result, Lesley spent her childhood being passed around among relatives. Sometimes she lived with her mother, and other times she stayed with her grandmother or one of her aunts. She never had much of a family life or sense of stability, and in early adolescence, she became extremely rebellious, leaving school at fourteen and hanging around at bars. Living from menial job to menial job, she was routinely fired for her irregular work habits. She was sixteen when she met John and barely knew him when she moved out of her grandmother's house to live with him. At first, they were homeless and slept in a boxcar parked on a railway siding. John was twenty-two and he understood her well; his upbringing had been nearly as tumultuous as hers. He also had a broken but not yet legally ended first marriage, as well as two daughters, one of whom was being raised by his stepsister and the other in an orphanage. He and Lesley felt that they were meant for each other.[56]

It may sound sentimental, but it really did turn out that their love for each other made all the difference in their lives. Within a couple of years of their meeting, John had found steady work driving a truck for the railroad company, retrieved his daughter Sharon from the orphanage, and divorced his estranged wife so he could marry Lesley. They had both longed for a stable family life, and Lesley, still in her teens, hoped to become pregnant as soon as they married. To her dismay, her fallopian tubes were blocked, and surgery to reopen them was unsuccessful. In 1976, her doctor, who had learned that Patrick Steptoe was seeking patients with tubal disease who were willing to undergo an experimental procedure, referred her to him. Lesley recalled that although Steptoe explained how in vitro fertilization worked, she did not understand what he was saying, nor did she realize that it had not yet succeeded in any other patients. She didn't care; whatever it was, she was willing to try. And then, to her joy, Lesley became pregnant on the first attempt.[57]

Meanwhile, more than 10,000 miles away, in the small Australian town of Churchill, Linda Reed and her husband, John, a teacher in primary school, were raising their two-year-old son Daniel and longing for another baby. Linda was twenty-four in 1978 and grateful for Daniel, particularly because her doctors had told her previously that they believed her tubes were too damaged from appendicitis for her to conceive. But soon after his birth, she had a series of operations for ovarian cysts, which caused further damage. When the couple read about the birth of Louise Brown, they consulted Linda's gynecologist, who referred them to the Melbourne IVF program.[58] Unlike the tiny group in Oldham—Steptoe, Edwards, and their research technician Jean Purdy—the Melbourne IVF team was large and impressive, boasting a number of prominent physicians and scientists based at two different hospitals. Linda Reed put her faith in Ian Johnston and his group. She became pregnant easily, although it is not clear whether, like Lesley Brown, she conceived on the first try. And just as Steptoe had with Lesley Brown, Johnston oversaw her care during pregnancy, delivering baby Candice on June 23, 1980. Linda was twenty-six years old.[59]

The next year brought success to Norfolk for the Joneses. Judith Carr was a twenty-eight-year-old fifth grade teacher in Massachusetts. She and her husband, Roger, an engineer, had been married since 1973. Judy Carr never had any trouble conceiving. She became pregnant three times, but each time the pregnancy was ectopic. In the end, she lost both of her fallopian tubes. Her gynecologist, learning that the Joneses were opening an IVF clinic in Norfolk, referred the couple to them, and the Carrs became one of fifty couples accepted for treatment that first year. Like Lesley Brown, she became pregnant on the first attempt.[60]

Carr was just one of the hundreds of thousands of women who suffered from tubal factor infertility in the United States. The Joneses would not be able to treat them all, and in any event, their clinic was soon joined by other programs. By the spring of 1982, there were seven programs altogether. The physicians from these programs met together informally, learning what each other was doing, discussing what was working and not working. They called on each other with questions and problems. The learning curve was steep, and the process seemed anything but orderly. Decisions about treatment—including what kinds of fertility drugs to employ and at what dose, or how many eggs to implant—resulted from trial and error. Patients were signing on for procedures at their own risk and their own cost. History seemed to be repeating itself. In the nineteenth century, when gynecologists began to treat infertility, the women who sought care from them were frequently both patients and research subjects. They paid for the opportunity to undergo experimental treatments that may or may not have helped them and may even have left them worse off than before. They underwent experimental surgery or took unproven drugs on the hope of bearing a child. In many ways, the development of IVF was simply a continuation of the nineteenth-century empiricism that advocates of scientific medicine had come to abjure in the twentieth century but that clinicians still often practiced.[61] Federally funded peer-reviewed research on in vitro fertilization could have changed this pattern. Such research would not only have promoted the development of agreed-upon protocols but also created a body of knowledge that would have provided the field with the data to understand and evaluate a range of treatments.

None of these things happened. Unwilling to allow the rest of the world to eclipse the United States, reproductive medicine specialists around the country felt a new urgency to move into the field now or risk falling behind, perhaps irretrievably. Howard and Georgeanna Jones provided them with a template. Patients who came to their clinic, based in a medical school and staffed by physicians affiliated with it, paid for treatment out of their own pockets. Once the Joneses stopped waiving their own fees, the cost of IVF rose to $5,000 per cycle, at a time when the median income in the United States was a little less than $18,000 a year. The business model of the Jones Clinic was replicated in most of the new IVF programs. Patients paid to undergo a brand-new treatment with a low rate of success, and at the same time they served as de facto research subjects. Marrs, who performed his early IVF procedures in a county hospital and refused to charge the patients until he earned some success, was an anomaly.[62] Each clinic created its own requirements for the kinds of patients they would accept into the program. In these years, nearly every program accepted only married couples, and most would only take women aged forty or younger. Many aspects of the new reproductive technology would change in the ensuing decades, but in many ways, the early programs created a model. A technique first used to help young women with tubal factor infertility fueled the creation of a market-driven enterprise that would touch nearly every aspect of reproductive medicine. The next two chapters chart that development.

4

FROM MIRACLE BIRTHS
TO MEDICAL MAINSTREAM

It was the spring of 1980, and Suzanne, an assistant professor in her early thirties, had just been awarded tenure and promotion at her university. She and her husband had wanted to start a family for the past few years, but they had been waiting until their careers were on a stable footing. Now they breathed a sigh of relief, bought a house, and made an appointment to see Dr. Luigi Mastroianni, the most prominent infertility specialist in the Philadelphia area and chair of the Department of Obstetrics and Gynecology at the University of Pennsylvania Medical School. "Why make that appointment now?" a friend asked. "You don't know if you'll have any trouble getting pregnant." True, Suzanne agreed, she had no reason to believe that she would have any difficulty, but there was a six-month wait to see Mastroianni. Just in case she did run into a problem, she told her friend, she was all set. As it happened, Suzanne conceived almost immediately and never needed the appointment, but such was the power of Mastroianni's reputation that only he would do. The other doctors in the department joked about "the Mastroianni magic." Couples would come in for their long-awaited appointment with Mastroianni, one of the younger doctors recalled, only to find that the wife was already pregnant. When they got the news, the couple would tell him, smilingly, "It was because of you, Dr. Mastroianni."[1]

It was an enviable reputation, one that Mastroianni could count on when he decided in early 1982 to create an in vitro fertilization program at Penn. Because he was considered a trustworthy medical voice in a city known for its veneration of tradition, his and the city's embrace of IVF in the 1980s augured well for the future of the technology. The program Mastroianni created was the seventh one to open in the United States, and it would become the sixth to have a birth.[2]

This chapter focuses on doctors and the programs they created in the 1980s, using Philadelphia as exemplar of the broader national trends that emerged in the early years of the development of this new reproductive technology. Across the country, the first successful programs were created at medical schools. Next came the community hospitals, some of them in partnership with private infertility practices. Freestanding, independent centers were the last to emerge. This is not a perfect chronology, but it does broadly represent how IVF developed in the United States as the technology made its way into the mainstream of reproductive medicine. In the next chapter, we move from medicine to society, examining the political and cultural conflicts that shaped the American response to the new technologies during this decade and thwarted efforts to regulate them.

Creating Programs, Raising Expectations

The fearlessness of Eastern Virginia Medical School in embracing the vision of the Joneses emboldened several of the more established academic medical centers to follow its lead. In some institutions, reluctant department chairs were pushed along by eager young fellows or faculty members. In others, including the University of Pennsylvania, prominent figures in the field took the lead. Everyone in the field of reproductive medicine was conscious of the advances in IVF being made in England and Australia. The international medical community took note when Steptoe and Edwards left the homey environs of Dr. Kershaw's Cottage Hospital in 1980 to move into an expansive new facility called Bourn Hall, Britain's first private IVF clinic, which soon began to recruit patients from around the world. Australia, with a population just slightly larger than the state of Texas, could boast of two prominent and successful in vitro fertilization centers in Melbourne, with a substantial staff of highly accomplished reproductive scientists and physicians. Because of their openness to doctors and scientists from around the world who were interested in learning from them, the Australians had become international leaders not only in treatment but also in training. Richard Marrs and Martin Quigley were just two among many who absorbed their lessons. Patients from across the globe filled their waiting lists.[3]

In the United States, young physicians created the early programs that followed the Joneses at Eastern Virginia—Marrs at the University of Southern California, Quigley at the University of Texas at Houston, and Alan DeCherney, a Mastroianni protégé who opened a clinic at Yale.[4] Older, more established reproductive specialists, often in positions of institutional responsibility, tended to be more cautious. Concerns remained—which proved to be unjustified—that the National Institutes of Health might take a punitive attitude toward other research programs if its grantees or their institutions became involved in human IVF.[5] In spite of such apprehensions, as 1981 came to a close with one birth in the United States, ten in Australia, four in England, and one expected in France in February, at least some senior physicians were willing to take the risk of ignoring any implied federal disapproval.[6] Anne Colston Wentz created Vanderbilt's program, which opened in February of 1982. At the same time, Mastroianni was laying the groundwork at the University of Pennsylvania for his program, which opened two months later, in April.

Luigi Mastroianni had served as John Rock's research fellow in the 1950s, and the two men developed a genuine rapport. Rock, whose only son had died in a car accident in 1946 at the age of twenty, thought of the younger man almost as a second son. Mastroianni returned his regard, even naming one of his sons John. A rising star in academic medicine, Mastroianni was barely forty when he came to the University of Pennsylvania Medical School in 1965 to chair the Department of Obstetrics and Gynecology. Shortly thereafter, he recruited Celso-Ramon Garcia, another of Rock's fellows. Years later, Mastroianni recalled with some pride that "contemporary reproductive medicine was developed" by the incumbents of those fellowships.[7] Garcia and Mastroianni were too young to have been involved in Rock's IVF experiments, which had ended before they arrived at the Free Hospital, but under the older man's tutelage they became expert tubal surgeons and reproductive specialists. Garcia, who had been deeply involved with Rock on the development of the oral contraceptive, directed Penn's Division of Human Reproduction. He was a talented surgeon, and his interests in the field of women's health ranged from infertility to contraception to sexuality in menopause.[8]

To Philadelphians, Mastroianni seemed like a pioneer when he decided to create an in vitro fertilization program, but he was actually something of a latecomer. An expert in IVF in nonhuman primates, in the 1970s he had not been ready to embrace its use in humans. As Steptoe and Edwards reached one milestone after another, Mastroianni continued to argue for additional primate research.[9] By the end of 1981, however, based on the health of the children born after IVF thus far, he was persuaded that the technology was safe. On January 18, 1982, he informed the legal counsel's office at the university's hospital that he intended to develop an IVF program. He inflated the number of healthy births around the world from fifteen—the actual number—to fifty, perhaps in error, perhaps thinking that fifteen babies would not have impressed the lawyers. He also informed them, accurately, that the American Fertility Society was about to announce its approval of IVF as a therapeutic option for women suffering from infertility. He also attached an informed consent document for the lawyers to review.[10]

Once he had decided on a course of action, Mastroianni moved swiftly. Within weeks, he had the necessary institutional approvals and made sure he had adequate staffing. The program began accepting patients in April. To direct the IVF clinic, he appointed thirty-three-year-old Richard Tureck, who had joined the faculty at Penn after completing his residency at Columbia in 1979.[11] Two other physicians, Luis Blasco and Steven Sondheimer, rounded out the early clinical team. Blasco, born and educated in Spain, had originally planned a career studying tropical diseases. While working in what was then called the Belgian Congo (now the Democratic Republic of Congo), he developed an interest in obstetrics and gynecology. In the late 1960s, with that country in turmoil, Blasco accepted an offer to move to the University of Pennsylvania, where he was drawn into the orbit of Garcia and Mastroianni.[12] Sondheimer, who was in his early thirties, had been both a resident and a fellow at Penn before he joined the department as a faculty member in 1980.

Almost as soon as the program opened, Sondheimer recalled, he had patients asking to participate. Even though not all patients were ideal candidates for in vitro fertilization, other treatments had failed them, and they believed that IVF offered their only chance, however slim, to have a baby.[13] At Penn, the reproductive specialists were integrating this new technology into their existing gynecological and infertility practices, which meant that such women were often long-standing patients. As Blasco told a reporter for the *Philadelphia Inquirer*, "A lot of [women] who have been our patients for five or ten years heard that we were starting up an in vitro program, and the personal pressure we felt was enormous. How could we tell them no?" Knowing these women's hopes and dreams, the doctors often agreed to bring them into the new program, even if they were on the cusp of forty or had multiple fertility problems.[14] More than three decades later, Sondheimer has not forgotten the women he treated in those early days. In fact, he doesn't have to remember some of them—they are still his patients.[15]

When the Penn IVF program began, the plan was to enroll thirty couples in the first year, with a long-term goal of two to three hundred a year.[16] Within six months, there were thirty-seven couples in the program. Mastroianni, concerned about cost, hoped to offer the technology at a reasonable price. In the end, Penn set the fee at $3,450, less than the Jones Clinic but still prohibitive for many. The program got off to a slow start. By February of 1983, just one pregnancy had been achieved. In hopes of increasing that number, the doctors established new criteria for acceptance into the program. Now, in order to be considered a candidate for IVF, a woman had to be under the age of thirty-nine with accessible ovaries; her husband had to have a normal sperm count.[17] With the new requirements, the fertilization rate began to improve even though individual doctors sometimes ignored these policies because they were unwilling to refuse IVF to their existing patients who did not meet the new criteria. By the end of 1983, 20 percent of fertilizations had resulted in a pregnancy, and there had been one birth. With practice, the numbers improved. In 1984, nine women had babies—eight single births and one set of twins. By the fall of that year, there were twenty-eight ongoing pregnancies.[18]

To judge from the procession of couples whose stories appeared in the local newspapers, Penn's patients were primarily white and middle class. Many of them chose to remain anonymous, so the fact that we saw no African American couples does not tell us for sure that they did not exist. But if they did, there could not have been very many, or we think the doctors we talked to would have remembered them. In other programs in Philadelphia and elsewhere, the situation was similar. Edward Wallach, who created Philadelphia's second IVF program, at Pennsylvania Hospital, recalled that most of its patients were middle-class couples with stay-at-home wives. Alan DeCherney described his patients at Yale as middle and upper-middle class. Only Richard Marrs in these early years seemed to have a broadly diverse patient base.[19] Given the high cost of treatment, even middle-class couples made significant financial sacrifices in their hopes for a baby. With no insurance coverage for IVF, some sold their homes. Others took out second mortgages or other loans. The doctors at Penn tried to save them money when they could, billing the insurance companies for whatever ancillary procedures that could be covered or deliberately choosing less expensive drugs if they thought they would be effective.[20]

With the birth of Jillian Elizabeth Johnston on September 23, 1983, the IVF program at Penn became the sixth in the United States to have a live birth.[21] Jillian's parents were fairly representative of the kinds of couples being treated in the program at the time. The Johnstons lived in southern New Jersey, not far from Philadelphia. Thirty-one-year-old Linda was a homemaker and her husband, Richard, forty-one, was a pipefitter. They were respectably middle income but not well-to-do. Neither were Lorraine and Charlie Twardowski, who had been trying to have a baby ever since they married in 1973. By the time they found their way to the University of Pennsylvania's IVF program, Lorraine had left her job to become a homemaker; her husband was a professor at Delaware Technical and Community College. Lorraine became Mastroianni's patient in 1982, and over the course of the following year, she underwent five cycles of IVF before experiencing a successful pregnancy. Their baby was born in March of 1984.[22]

Learning by Doing

Recalling his early experiences performing IVF in the early 1980s, Steven Sondheimer told us with some bemusement, "When I think about those days, I'm amazed that anybody got pregnant with our treatment. But they did. Some people . . . got pregnant twice. It's humbling to think about. And of course, Steptoe and Edwards are the greatest examples of how, with rudimentary techniques, you can have success." He paused for a moment, then added, "Not as much success as you have today."[23] He and the other doctors engaged in this new endeavor across the country and around the world were feeling their way, and all of them found that it helped to discuss protocols and techniques with others in similar situations. There were meetings and conferences for physicians who were offering IVF. Sometimes, less experienced teams would call on those with more expertise for advice. Esther Eisenberg, who directed the IVF program at Pennsylvania Hospital in the mid-1980s, remembered "how frustrated everyone was that after we got our first pregnancy" no more followed. "We had problems and we couldn't figure out what was going on," she said, so they invited embryologist Jacques Cohen, who worked with Robert Edwards, to "come from New York . . . to review our whole program." When Cohen observed the embryologist shaking the vial containing the eggs, he saw that some powder on the glove was getting into the tube. As it turned out, that tiny bit of powder "was enough to interfere with the embryo development." Until Cohen noticed, they had no idea such a simple thing was derailing their success.[24]

Even the instruments and equipment had to be built from scratch. Richard Marrs recalled that when he began his IVF program, with its less than shoestring budget, there were no instruments and no materials. So "I went down to the machine shop and talked to one of the machinists and sat with him and [we] made our first aspiration needle that I could use laparoscopically to aspirate a follicle, because we didn't have needles to buy." To the machinist's question, "How do you know what you need?" Marrs answered, "Well, I know what I don't need because these regular beveled needles are too long, too sharp and they're too traumatic for the ovaries so . . . I want to shave down the bevel, I want to blunt it, I want to do this, do that and so we went back and forth and made our first needles and the same with our transfer catheters. I worked with [the machinist]," he said, because after all, "Somebody had to do that because there was no availability of anything as far as instrumentation."[25] Yes, "somebody" at new programs everywhere needed to jury-rig equipment, figure out what culture medium to use, repurpose an instrument or build one from scratch.

Standardized instruments, culture medium, and equipment were in the field's future, of course, but back then, all these elements varied from center to center. Embryologists and their lab staffs everywhere made their own culture medium, and everyone wondered about proper temperature and light levels for the rooms in which the eggs were fertilized. All these variations notwithstanding, there were a few essential components of a successful program in these years. The first requirement was outstanding surgical skills. Because eggs were typically retrieved by laparoscopy, a doctor's skill in performing that technique could make the difference between success and failure. Alan DeCherney, who had received his training at Penn, recalled that the department of obstetrics and gynecology under Mastroianni and Garcia was among the first in the United States to make extensive use of laparoscopy, long before the advent of in vitro fertilization. Everyone who trained or practiced in that department knew how to perform a laparoscopy and do it well; every one of the infertility specialists were surgeons as well as reproductive endocrinologists. Some doctors, if they discovered that a patient's eggs were inaccessible using the minimally invasive laparoscopic approach, would even perform a laparotomy to retrieve them. Although its use was rare for this purpose, Sondheimer recalled that in some cases opening the abdomen was the only way to find and remove the ripened eggs. In at least two instances, laparotomy made it possible for his patients to have babies who would never have been conceived otherwise.[26]

IVF Comes to the Nation's Oldest Hospital

The University of Pennsylvania and its medical school were, and still are, located in an area of the city called West Philadelphia, about a mile and a half from what Philadelphians call "center city," by which they mean the downtown. Near the heart of center city is Pennsylvania Hospital, which opened the city's second IVF program in 1983. The nation's oldest hospital, founded in 1751 by Benjamin Franklin and local physician Thomas Bond, Pennsylvania Hospital had well-regarded residency training programs and a prominent maternity unit. Many of its physicians enjoyed national reputations. Some of them held faculty appointments at the University of Pennsylvania's medical school, although in the 1980s the two institutions were separate entities. (In 1997, Pennsylvania Hospital was acquired by Penn and is now a part of the Penn Health system.) The IVF program at Pennsylvania Hospital, founded by Edward Wallach, represented the second stage of the development of IVF in the United States as the technology moved beyond medical schools and into community hospitals. Wallach had come to Philadelphia in 1965 as a faculty member at Penn's medical school, and like his mentor, Celso Ramon Garcia, he had a reputation for outstanding surgical skills.[27]

Wallach had a reputation for diplomacy, a characteristic he shared with Mastroianni, his former chair. The diplomatic skills of these two men are the likeliest explanation for the fact that neither of these IVF programs generated controversy in the community, where, given Philadelphia's strong Catholic presence, some opposition might have been expected. In the 1980s, more than a third of the city's population was Catholic, and the leader of the Philadelphia Archdiocese was the powerful and theologically conservative John Cardinal Krol. Wallach made an effort to forestall opposition from the Church, recalling that in advance of announcing his program, he sought an appointment with an important diocesan official who sat on Pennsylvania Hospital's Board of Managers, where he explained his department's plans. The meeting was clearly successful, because no objection came from the diocese.[28] Mastroianni himself had muted potential objections from the Church by transferring into the uterus all of a patient's eggs that appeared to have fertilized acceptably. As he told a reporter, he had no intention of "waging war against the moralists." Not discarding any fertilized eggs meant that he could "eliminate the discussion of what to do with the rest" of them, he said, because there were none left over.[29]

Wallach appointed Esther Eisenberg to direct the new program at Pennsylvania Hospital. She had completed her fellowship in reproductive endocrinology and infertility with him the year before, after which she joined his practice as a partner. Developing IVF, she told us, "was Ed's dream" and "his drive and vision" made it possible, but her contributions were also significant. "I can make things happen," she said, and so she did, including the retrieval of Wallach's first patient's eggs. "It was our first retrieval," she recalled, "and we're in the [operating room]. Ed was puncturing the ovary, trying to get the egg. He couldn't get an egg, so he calls me, 'Esther, come down here.' And I go and get the egg."[30] Pennsylvania Hospital had its first birth on May 22, 1984. Wallach told us in 2016 that every year on that day he talks to the mother of his first IVF baby. Like Sondheimer and others among these early practitioners of IVF, Wallach felt a long-term interest and investment in the women he treated.[31]

Wallach was also an important mentor to the next generation. He recognized talent in young physicians, and he was committed to helping them develop. In a field that was—with a few exceptions, such as Georgeanna Jones and Anne Colston Wentz—largely a man's world in the 1980s, Wallach nurtured and supported the talents of women like Eisenberg and served as a mentor to PonJola Coney, known to her friends as PJ, who we believe was the only African American among this early group of IVF pioneers. Coney grew up on a farm in Mississippi and was the first in her family to graduate from college. Despite the advice of her professors at Xavier University, who urged her to apply to medical school, she chose to become a medical technologist because she felt the need to go to work right away in order to help her family. Returning to Mississippi after graduation, she took a job at the acute care laboratory in the medical school of the University of Mississippi, where she came to know some of the physicians. Impressed by her abilities, one of them urged her in the strongest terms to become a doctor. She finally listened. "Based on that conversation," she recalled, she applied to and was accepted at the University of Mississippi Medical School, graduating in the late 1970s. She chose obstetrics and gynecology as her field, partly because she enjoyed the work but also because the faculty in that field "treated me with decency. And that wasn't necessarily the case in the other specialties."[32]

After medical school, Coney completed her residency at the University of North Carolina. While she was there, her work brought her to the attention of Luther Talbert, head of the reproductive endocrinology and infertility division. He became a mentor, and when she told him she wanted to become a subspecialist in his field, he said he would help her find a fellowship. To her great disappointment, she remembered, she didn't get a single response from any of the programs to which she applied. She did not tell us outright that the lack of interest in her had to do with race, but it is impossible to imagine a white man or woman who had been mentored by a leading figure in the field not getting at least one interview. "It was terrible," she told us. "I went to [Dr. Talbert] and I said, 'I guess that dream is down the tubes. It's not going to happen.'" Talbert, however, was not about to let a promising career collapse before it could even start. He called Wallach at Pennsylvania Hospital. Wallach was getting ready to decide to whom he wanted to offer a fellowship and asked Coney if she could come to Philadelphia the next day. She could and she did.

A week later, he offered her the fellowship. Coney recalled, "I said to myself, 'I know that's going to be a tough one for him, because he's going to be the first one to take an African American fellow. A female.' I could never thank him enough." Coney had been a standout resident who faced one obstacle after another just to earn the right to have the chance to prove herself. She was grateful to Wallach for giving her an opportunity to succeed.[33] She arrived at Pennsylvania Hospital in 1982, and the following year, she "was part of the startup [of the IVF program at Pennsylvania Hospital] with Ed and Esther." After completing her fellowship, she went on to direct IVF programs in the 1980s at two academic medical centers in the Midwest—both the first in their respective states.[34] Wallach, whom we interviewed at some length, spoke of both Eisenberg and Coney with considerable pride, but he never mentioned the story that Coney told us. Perhaps he simply thought he was doing his job when he nurtured the careers of these talented women, but they have never forgotten how his support made an important difference at a critical point in their careers.

The University of Pennsylvania and Pennsylvania Hospital dominated the IVF landscape in the region for the remainder of the decade. Wallach left Philadelphia for a position at Johns Hopkins Medical School in 1984, after which Pennsylvania Hospital's program moved toward a private practice model. The ways in which the two institutions diverged in their trajectories over the course of the 1980s allows us to explore the different ways in which the new technologies developed as IVF moved from a medical curiosity to a mainstream infertility treatment.[35] At the University of Pennsylvania, the IVF program remained remarkably stable throughout the decade. By 1988, it had six board-certified endocrinologists, all experienced clinicians who had been with the program since its early years. Like other IVF clinics at academic medical centers, it attracted a mix of patients, ranging from young women with tubal disease to couples who had tried and failed in other programs or were rejected elsewhere because of age or diagnosis and were coming to Penn for one more chance, maybe their last.

From 37 couples in its first year of operation, the program had grown to 196 in 1988. Penn clearly had not held the line in enforcing age restrictions, or perhaps the doctors made a decision to expand them. In that year, 25 percent of the patients were women age forty or older. The most common diagnoses of those patients, however, were still what they had been in 1982. About 40 to 50 percent of the women had fallopian tube disease, and 20 to 30 percent suffered from endometriosis. Some patients had both. Ten percent of the couples had a diagnosis of unexplained infertility, and another 10 percent had immunological infertility. Just 6 percent of couples in the Penn program suffered from male infertility either with or without other conditions.[36] Success rates had improved for the program. In 1987, 12.5 percent of the couples in the program took home a baby. By 1988, it was almost 20 percent, and the younger the woman, the better the outcome: couples in which the wife was under forty and who had no male factor infertility had a take-home baby rate close to 22 percent.[37]

By 1988, Pennsylvania Hospital's IVF program had partnered with a private practice, that of infertility specialist Stephen Corson. Its program served more than four hundred patients in 1988. Its patients were younger than those being treated at Penn. Just 4 percent of Pennsylvania Hospital's IVF patients were over age forty in 1988, and a greater number of the patients overall, 53 percent, suffered from tubal disease. Like Penn's program, it was well staffed with six reproductive endocrinologists. Its success rates were comparable; they were slightly higher than the University of Pennsylvania's in 1987 and slightly lower in 1988.

Corson, who directed the program, was an advocate of using IVF to treat male infertility even in the absence of other diagnoses, a controversial practice given that there were no data to suggest that it worked. His program also made significant use of GIFT, short for gamete intrafallopian transfer, a procedure in which eggs were retrieved just as they were for IVF, but instead of being fertilized externally, the egg and sperm were transferred into the woman's fallopian tube to be fertilized. GIFT is rarely used now, but it was popular in the 1980s, and Pennsylvania Hospital's program used it for patients with endometriosis as well as for male-factor and unexplained infertility.[38]

By mid-decade, the University of Pennsylvania and Pennsylvania Hospital no longer housed the only programs in the region.[39] IVF clinics at Albert Einstein Hospital in Philadelphia and St. Luke's Hospital in Bethlehem opened in 1985 and 1986, respectively, and several private IVF practices were started later in the decade. By 1988, the region had a total of nine IVF programs, and *Philadelphia Magazine* called the area "one of the nation's hot spots for fertility specialists."[40]

Throughout the country, the numbers of programs were growing, and so were the indications for which IVF was used and sometimes overused. When PonJola Coney created that state's first IVF Program at the University of Oklahoma, she recommended the technology only for patients who failed to become pregnant with other, less invasive medical treatments, and she followed the same course of action later in the decade when she founded Nebraska's first IVF program. Because she had considerable success with medical treatments, it turned out that the number of her patients who needed IVF to conceive was relatively small compared to the patients of some of her colleagues. " 'Where are you finding all these people who need IVF?' " she remembered asking them, curious about the large number of IVF cases, "particularly in small markets." Her colleagues' response? " 'Well, they're infertile, aren't they?' " It appeared that "once they had the laboratory and staff set up everyone went to IVF." Meanwhile, "There I was," she said, "still getting most of my patients pregnant by other means."[41]

In the latter half of the 1980s, advances in ultrasound-guided egg retrieval transformed the first stage of IVF, the removal of eggs for fertilization. In

1981, Danish gynecologist Susan Lenz began retrieving eggs transvesically, by inserting a needle through the bladder into the ovaries under abdominal ultrasound guidance.[42] Although this procedure did not involve surgery, it did often require general anesthesia, was uncomfortable, and could be painful. In the United States, it was not universally adopted.[43] At mid-decade, however, two improvements transformed the procedure. The first was the shift from transvesical to transvaginal egg retrieval. Now the physician could insert the needle through the vagina into the ovary to retrieve the eggs, which was less painful for the patient than going through the bladder. Two different research teams—one in France, the other in Sweden—first reported on the new technique in 1985.[44]

Lenz herself was among those who developed expertise in this method, and she taught it to others, spending three months in Melbourne at mid-decade training doctors in the newest techniques.[45] In 1986, the procedure became even simpler with the introduction of the vaginal ultrasound probe. Transvaginal ultrasound-guided egg retrieval simplified the process of removing eggs for fertilization and could be performed under local anesthesia.[46] The new technique soon became standard for every type of program in the United States—whether academic, hospital based, or private. It also helped promote the growth of freestanding centers, because ultrasound-guided transvaginal egg retrieval did not require a surgeon or an operating room.

Professional Challenges

In vitro fertilization programs burgeoned nationwide. Thirty-seven clinics opened across the United States in 1983, and another thirty-one opened in 1984. By 1988, there were nearly two hundred clinics treating about ten thousand patients.[47] Forty-two states had at least one IVF provider, and several states had many more. California, which still has more IVF centers than any other state, had twenty-four IVF programs; Texas, fifteen; and New York, twelve. Pennsylvania and Florida each had nine. Alaska, Montana, and Idaho had no IVF clinics of their own, but their residents could gain access to the new technology through a partnership between gynecologists in their states and the University of Washington's medical school in Seattle. Alaska and Idaho each had one collaborating practice with the Seattle medical school. Montana had five. Specialists at the medical school trained gynecologists at the satellite locations to initiate and monitor cycles, and the University of Washington physicians provided ongoing advice and support. The patients then traveled to Seattle for egg recovery, fertilization, and embryo transfer.[48] No other states without their own clinics had similar arrangements. If couples in North and South Dakota, Wyoming, Maine, or Arkansas wanted to access IVF treatment, they had to travel to another state.[49]

By 1988, a growing number of freestanding clinics run by private practitioners had joined the academic medical centers and hospital-based practices. With fewer institutional constraints in the form of institutional review boards and ethics committees, and the advantages of flexibility in pricing, private centers would become increasingly important in the next decade. Their growth was already evident by the late 1980s.[50] In 1983, just 11 percent (a total of five) of the approximately forty-four clinics then in existence were non-hospital-based private practices; by the end of 1988, that figure had risen to 32 percent. Still, most programs at this time remained at medical schools or hospitals: 38 percent were at academic medical centers, and another 30 percent were hospital-based practices (some of them in collaboration with private practices).[51]

The organized reproductive medicine community, as represented by the American Fertility Society, embraced the treatment options made possible by the new reproductive technologies. Its members were, however, worried about how to make sure that those who offered the treatment were qualified to do so. A doctor did not have to be a board certified reproductive endocrinologist, or even an obstetrician/gynecologist, to open an IVF clinic. Any licensed physician, regardless of training, expertise, or skill, could do so. There were no regulations. By 1988, highly experienced reproductive endocrinologists and embryologists in established programs were achieving birth rates of 20 percent and even slightly higher, especially if they were treating young women with fertile husbands. Nevertheless, fully 21 percent of the clinics in operation at that time did not have a single live birth in 1987, and the national "take home" baby rate was just 9 percent.[52] Even as the experts in reproductive medicine who led the American Fertility Society were attempting to set and encourage professional standards in IVF practice, no one had to follow them. Individual practitioners could pretty much make up their own rules.

In an effort to exert some oversight over the new technologies, the American Fertility Society had developed a set of guidelines, including minimum standards for IVF clinics, in 1984. It also created a new organization within the larger group, the Society for Assisted Reproductive Technology (SART), with rules governing admission to membership. To become members of SART, IVF providers were required to report their results to the society annually. The data were anonymized, with no information reported on individual clinics, but the overall results were summarized and made public in the parent society's journal *Fertility and Sterility*. This registry could do nothing to stop anyone from opening an IVF clinic, but it did restrict membership in SART to clinics that followed its guidelines.[53]

The leaders of the field were not blind to the multiple dilemmas that the new reproductive technologies created. Testifying at two congressional hearings, in 1988 and 1989, these experts were clearly worried about inexperienced—or rogue—practitioners who could tarnish the reputation of the entire field out of either ignorance of or indifference to professional standards. "There is always somebody . . . getting ready to open a new program," Benjamin Younger, president of the American Fertility Society, told the members of the Subcommittee on Regulation and Business Opportunities of the Committee on Small Business. "Some of [them] do not meet [our] minimum standards." These practitioners wanted to open such programs, he concluded ruefully, "Because it sounds easy."[54] It was unclear, however, what the society could do about the situation. In a new field, with virtually no oversight or regulation, it should come as no surprise that some doctors cut corners, duplicitously advertised inflated and even invented success rates, or committed outright fraud.

Reproductive medicine was becoming a lucrative area of medical practice. Without regulation and without insurance coverage, either of which might have put some brake on the fees that some clinics charged, the new reproductive technologies had the potential to become a free-market bonanza for some.[55] "The entrance of for-profit organizations" into medical practice, Alan DeCherney testified at one hearing, had the potential to open "doors for economic exploitation." Infertility patients, he said, were "exceptionally vulnerable." Because "the practice of infertility is safe," he continued, and success rates were low even in expert hands, it was harder for patients to prove they had been duped. DeCherney believed nevertheless that the number of doctors exploiting their patients was "proportionately quite small."[56] Richard Marrs was not so sure. He had been involved in developing guidelines for SART membership, and he wondered whether the reproductive medicine community would be able to police itself. The guidelines on which he and others had worked so hard to develop were met, he said, with open hostility from the "rank and file practitioner" who believed that anyone who wanted to do it should be allowed to open an IVF clinic and be admitted to membership in SART. Although he knew his opinion was a minority one, he said he believed that federal "regulatory control" of the new technologies was warranted.[57]

How Do We Know If IVF Will Help Us?

As IVF clinics proliferated, it became increasingly difficult for patients to know how to choose one. They could ask a clinic or doctor directly about success rates on the assumption that the higher the success rate, the better the program. But what if one infertility practice had a great deal of success in getting its patients pregnant with more conventional treatment and did not recommend IVF unless those methods failed, and another program started everyone on IVF right away? As Coney said of the first program she directed, "We would use conventional techniques first. We would move the patient to IVF if other interventions failed." This was common practice at medical school–based clinics more generally, she said, which "may explain why the pregnancy rates were somewhat lower in academic programs initially, because they were mostly purists."[58] Other factors influencing success rates included the expertise of a program's physicians and embryologists, but there were also factors beyond anyone's control, such as a woman's age and a couple's diagnosis.

All other things being equal, in the late 1980s, an IVF program treating women who were age thirty-five or younger, whose husbands (at the time, most clinics treated only married heterosexual couples) had a normal sperm count, and whose only diagnosis was tubal disease, had a higher success rate than an equally skilled practice with a greater number of older patients or couples with more complex diagnoses. (In fact, with some emendations— fertility centers welcome gay and lesbian couples and single women nowadays, and overall success rates are higher—this generalization is still true. The younger a woman is, and the more straightforward her problem, the more likely she is to become pregnant.) Some clinics in the 1980s used IVF almost exclusively for fallopian tube disease. In 1988, more than 90 percent of Richard Marrs's patients, for example, were in that category, and he was not alone.

By the end of the 1980s, however, the uses of IVF had expanded in three ways. The first was in diagnosis. Our sample of twenty-four IVF programs in 1987 and 1988 showed that 70 percent listed tubal factor infertility as the most common diagnosis for their IVF patients, but clinics were also employing the technology to treat endometriosis and unexplained infertility. About 40 percent of clinics said that endometriosis was a primary diagnosis in one-fifth or more of their patients. In addition, using IVF for male infertility, even in the absence of any condition in the wife, was on the rise. In 1987, about a quarter of these clinics reported that 20 percent or more of their patients suffered from male factor infertility, and just a year later that figure had risen to more than a third.[59] Why use IVF for such conditions? Endometriosis, even if it does not block the tubes, can cause adhesions, making it hard for a newly fertilized embryo to make its way into the uterus for implantation, so in vitro fertilization seems a reasonable treatment. But employing it for male factor infertility was different. Some doctors believed that IVF improved the likelihood of sperm penetration. There was one technique that involved piercing the shell of the egg (the zona pellucida) before it was fertilized to make it easier, at least in theory, for the sperm to enter the egg's interior. Another used "micro-injection" of sperm to achieve the same goal. Neither worked, as studies soon showed, but for a while they were employed fairly extensively.[60]

The second way that IVF expanded was by age. Early on, because the first successful IVF births were to relatively young women, the guidelines of many programs set an upper age limit somewhere between thirty-five and thirty-nine, but by the second half of the decade, many clinics were willing to take patients over age forty, with most of them setting an upper limit of forty-three. Women in their early forties, however, accounted for only around 5 to 10 percent of patients in 1988. A few programs accepted some women over forty-five, but there were never more than one or two such patients in any of those practices.[61]

The third was the introduction of donor eggs and embryos and gestational surrogacy. As word of pregnancies from donor eggs rippled through the media, clinic directors had to decide whether to incorporate them, or donor embryos, into their practices. In the early years of IVF, when a woman had to undergo a laparoscopy for egg retrieval, there were almost no paid egg donors. Some clinics did use altruistic ones, however, often family members, and there were a few offering what were called "egg-sharing" programs.[62] The latter practice was more common in Australia, whose doctors may have been the first to ask their patients to donate their unused eggs altruistically. Melbourne IVF pioneer John Leeton was occasionally asked how a donor felt if she failed to conceive but the person to whom she donated one of her unused eggs did become pregnant. So long as "anonymity between donors and recipients was guaranteed," he said, and both couples received appropriate counseling, he did not see a problem.[63] In fact, Leeton's first donor egg pregnancy had presented that very situation. In 1982, the recipient of a donor egg conceived but the donor did not. According to Leeton, "this sensitive situation was readily accepted by the donor."[64] He may have been correct, but the recipient miscarried at eleven weeks, so there was never a birth. We cannot help but wonder whether the donor would have had a different reaction if the pregnancy had produced a baby. When Leeton's Melbourne IVF group did achieve the world's first birth using a donor egg in November 1983, there was no record of what that donor thought. It is not clear if she even knew that her egg was the one that allowed a twenty-five-year-old Italian woman with no ovarian function to become a mother.[65]

In the United States, only a few IVF clinics had egg-sharing programs similar to the ones in Australia. Among those who did, the most prominent was the Jones Institute, where "generous, consenting IVF patients having extra eggs . . . provided [them] anonymously for fertilization." The Jones Institute also used eggs donated by friends or relatives of the patient being treated, and by 1987, it had thirty children born from donor eggs.[66] The practice of sharing a woman's unused eggs with another patient was never common in this country, and once embryo cryopreservation became routine, egg sharing became even less so.

The first birth from a frozen embryo occurred in the Netherlands in 1984, and over the next couple of years, about two dozen babies who began their existence as frozen embryos were born in Australia, Israel, and several countries in Europe. The first birth from a frozen embryo in the United States was to a patient of Richard Marrs. By the time she delivered the baby on June 4, 1986, there were three other frozen embryo pregnancies in the United States (one of them another patient of Marrs).[67] Embryo cryopreservation gave women another choice of what to do with their extra eggs. Now, she could save them and have them fertilized for her own later use.[68]

By the end of the 1980s, an estimated five hundred women around the world had used donor eggs of one sort or another to achieve a pregnancy and birth, most of them young women without functioning ovaries. But their use was still controversial. In 1988, only about 13 percent of American IVF clinics had donor egg programs. Luigi Mastroianni was initially reluctant, but in 1989 he decided to allow their use at Penn.[69] In addition to egg donation programs, the first gestational surrogacy program was founded at mid-decade, offered by the IVF program at Mt. Sinai Hospital in Cleveland under the direction of Wulf Utian.[70]

There were other changes as well that began to shape the modern contours of the field. The number of private, for-profit programs increased. IVF success rates gradually improved, especially among the more established clinics with skilled embryologists and physicians. Protocols became more standardized. Couples seeking IVF, however, remained on their own in terms of choosing a program. Public scandals were uncommon, but in reality it would be almost impossible for couples to know if they were being cheated or for legal authorities to know whether clinics were violating any laws. Prospective patients asked, "How many women treated end up with a baby?" Clinics, in contrast, wanted to define success by numbers of eggs fertilized, or number of cycles that resulted in a pregnancy. Sometimes IVF programs did not even talk about births, just pregnancies, but as every woman who ever had a miscarriage knew, a pregnancy does not always end in a birth.[71]

The official stance of the American Fertility Society was that the profession itself, and not the federal government, should set standards and create guidelines for practice, although there were some dissenters who believed that federal regulations were essential. There were some lawmakers who agreed, but the politics of the 1980s militated against regulation just as a new pronatalism was increasing the demand for reproductive services. How the intertwining of these two factors determined national policy on the new reproductive technologies during the 1980s is the focus chapter 5.

Figure 1. John Rock, sometime in the 1940s. A prominent infertility specialist in Boston with a growing national reputation as a researcher and clinician, Rock became a national media sensation in 1944, after he and his research assistant, Miriam Menkin, reported the first successful in vitro fertilization of human eggs. Courtesy of the family of John Rock.

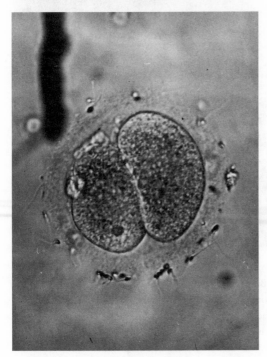

Figure 2. Original photograph of one of the first human eggs reported to have been fertilized in vitro, 1944. This egg, in the two-cell stage, was one of four eggs fertilized in that year by John Rock and Miriam Menkin. Courtesy of the Countway Library of Medicine, Harvard University, and the family of John Rock.

Figure 3. Miriam Menkin around 1935, with her husband, Valy Menkin. This photograph was taken about three years before Menkin began working for John Rock. Courtesy of the Countway Library of Medicine, Harvard University.

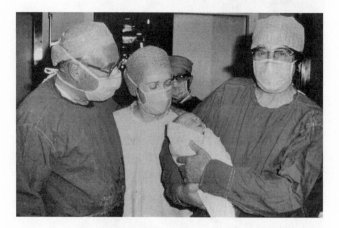

Figure 4. *Left to right:* Patrick Steptoe, Jean Purdy, and Robert Edwards with Louise Brown, the first baby to be born as a result of in vitro fertilization, shortly after her birth on July 25, 1978. Purdy's presence in this photograph is no accident. She was Edwards's "indispensable" research technician. He and Steptoe considered her contributions to be critical to their success. Courtesy of Bourn Hall Clinic.

Figure 5. Patrick Steptoe (*left*) and Robert Edwards (*right*) at a press conference in January of 1979. By this time, they had achieved two births using in vitro fertilization: Louise Brown on July 25, 1978, and Alastair MacDonald on January 14, 1979. Courtesy of Bourn Hall Clinic.

Figure 6. First Bourn Hall Meeting, 1981. Hosted by Patrick Steptoe and Robert Edwards, the meeting, held September 3–5, brought together some of the most important researchers and practitioners working on in vitro fertilization from several countries. Seated in the front row (*left to right*) are Robert Edwards, Jean Purdy, and Patrick Steptoe. Standing in the row behind them are Howard and Georgeanna Jones (*third and fourth from right*). The Australian researchers present were Alan Trounson (*first standing row, second from left*), John Leeton (*back row, right*), and directly in front of him (*left to right*) Alex Lopata, Andrew Speirs, and Ian Johnston. The photograph is said to have been taken by famed Swedish photographer Lennart Nilsson, who was attending the meeting. Courtesy of Bourn Hall Clinic.

Figure 7. Ron Wyden, Democratic member of Congress from Oregon, in 1989. Wyden's bill, the Fertility Clinic Success Rate and Certification Act, was the only national legislation addressing the new reproductive technologies to become law. Passed by voice vote and signed into law in 1992, Wyden's bill gained bipartisan support by shifting the paradigm from regulation to consumer protection. This law created a mechanism for IVF clinics to provide information on success rates to potential patients. It did not take any steps to regulate the technology itself. Courtesy of the Library of Congress, Prints and Photographs Division, CQ Roll Call Photograph Collection.

Figure 8. Anne Colston Wentz (*center*). Wentz directed the in vitro fertilization program at Vanderbilt University, which in early 1983 became the fourth in the United States to have a baby as a result of the technology. This photograph is from the 1980s. Wentz had been a fellow under Georgeanna Seegar Jones (*left*). Howard Jones is at right. Courtesy of History of Medicine Collections, Vanderbilt University Libraries.

Figure 9. PonJola Coney, the only African American woman among the IVF pioneers, when she was a fellow at Pennsylvania Hospital in Philadelphia, 1983. Coney went on to create the first in vitro fertilization programs in Oklahoma and Nebraska. She later became a leading figure in academic medicine, serving as a medical school dean and vice president. Courtesy of Pennsylvania Hospital Historical Collections.

Figure 10. Edward Wallach (*right*) with Howard and Georgeanna Jones. Wallach created Philadelphia's second in vitro fertilization program, at Pennsylvania Hospital, before moving to Johns Hopkins University, where he enjoyed a long and productive career. He served as a mentor to PonJola Coney, Esther Eisenberg, and other young physicians entering the field. Courtesy of the Alan Mason Chesney Medical Archives of the Johns Hopkins Medical Institutions.

5

THE ELUSIVE SEARCH
FOR NATIONAL CONSENSUS

The United States suffers from an epidemic of infertility, IVF expert Martin Quigley told *Time* magazine in 1984. According to the magazine, the rate of infertility in the United States had tripled over the past twenty years, and the childbearing potential of American women was in jeopardy. "Doctors place much of the blame" for the "epidemic," wrote reporter Otto Friedrich, on "liberalized sexual attitudes," resulting in an increase in pelvic inflammatory disease, delayed childbearing, and the stress of a career. "Women executives" were considered particularly susceptible to stress-induced missed periods and subsequent fertility problems.[1] If an infertility "epidemic" wasn't enough to worry about, American women also faced a "man shortage," which was hitting women with career ambitions particularly hard. *Newsweek* delivered the "traumatic news" in 1986, claiming that "dire statistics confirmed" a finding that "everybody suspected all along": high-achieving, educated women who were "still single at 30" had only a 20 percent chance of finding a husband. And if they were unable to find one by the time they turned forty, they were almost surely doomed to permanent spinsterhood. In the article's most famous line, single women over forty were "more likely to be killed by a terrorist" than to marry, with "a minuscule 2.6 percent probability of tying the knot."[2]

These alarming assertions became baked into the culture of the 1980s. Neither was true. The *Newsweek* story was so egregiously false that the magazine took the rare step twenty years later of retracting the entire article.[3] But both of these claims struck a familiar chord, warning young women that if they chose to make their work a priority they would never find a husband, or if they did manage to marry, the stress from their careers would make them infertile.[4] After a decade of legal, professional, and educational gains for women, the forces of reaction were setting in. What better way to undermine those gains than to tell young women that their ambitions would lead to a lonely and sterile middle age? But as wrong on the facts as they were, the articles and books and the reams of commentary they generated struck a powerful emotional chord in the 1980s. This was a decade of deeply contested ideas about gender roles and sexuality. The fears evoked by these claims helped frame the public's response to the new reproductive technologies, shaping the particular—and uniquely American—ways in which these technologies emerged, expanded, and were received during the decade.

Four interacting forces drove this process. The first was a new wave of pronatalism acting in concert with a simmering backlash against feminism. Societal pressure to procreate merged with a call for women to scale back their professional ambitions to focus on their families. The second was a significant increase in demand for infertility treatment, arising in part from the widespread publicity that accompanied the first IVF babies born in various regions of the country. It would have been almost impossible for any newspaper reader to miss these stories, which not only highlighted the successes of the new technology but also showed readers where to find treatment if they were having difficulty conceiving. Third, the first two developments were undergirded by the rise of conservatism during the two terms of Ronald Reagan's presidency. And finally, there was widespread media coverage of both the lack of regulation and the high cost of the new reproductive technologies. Democrats, who controlled the House of Representatives but not the Senate, came to believe that these new ways of making babies should be regulated at the national level. Because government was divided, with Republicans controlling the Senate and the White House, Democrats sought ways to find common ground. That proved difficult. A total of five congressional hearings were held. Four of them failed utterly. The fifth did indeed result in legislation, but only because it reframed the legislative issues from regulation to consumer protection.

Pronatalism in Action

As we noted in chapter 3, by 1980, stories celebrating the pleasures of a life without children, so prevalent in the prior decade, had virtually disappeared from the nation's magazines and newspapers, replaced by reports like the ones that opened this chapter. Marriage and parenthood were back in fashion in the media and in real life, and the children of the baby boom who were now in their mid- to late twenties were marrying and starting families. They were joined by their slightly older counterparts who, having waited even longer to settle down, were now tying the knot and decorating their nurseries. The nation's birth rate, which had reached a low of 14.6 births per thousand women in 1976, began going up, to 15.9 in 1980 and 16.7 in 1990.[5] Even more of a bellwether of changing behaviors than the overall increase in the birth rate was the rise in the age at which women had their first child. The number of women having their first births in their late twenties (instead of at a younger age) rose from 38.2 per thousand women in 1980 to 44.1 in 1990. Even more striking was the increase in the first birth rate for women aged thirty to thirty-four, which rose from 12.8 in 1980 to 21.2 in 1990; among women aged thirty-five to thirty-nine, the first birth numbers went up even more dramatically, from 2.6 per thousand women in 1980 to 6.7 in 1990.[6]

Many of these women expected to have careers—or at least jobs—as well as families. Practically speaking, a comfortable middle-class standard of living still often required a two-earner household; although the high inflation rates of the late 1970s and early 1980s had begun to ease, wages remained stagnant.[7] Furthermore, women who had gone to college and perhaps to graduate, medical, or law school in the 1970s and early 1980s, when the feminist movement was at its peak, expected to have professional careers, just like men. But by the mid-1980s, after a decade or so of gains for women under the law, progress ground to a halt. Anti-feminism, masquerading as post-feminism or even as a new version of feminism, promoted the message that women's empowerment could be displayed by staying at home and being supported by a man.[8] Studies showed that by the end of the 1980s, men had become "less likely than women to support equal roles for the sexes in business and government, less likely to support the Equal Rights Amendment—and more likely to say they preferred the 'traditional family' where the wife stayed home."[9] This backlash—the word became the title of journalist Susan Faludi's book on the subject—set back the movement for gender equality and cleverly co-opted the language of women's rights to promote an anti-feminist agenda.[10] Career-minded women were told that if they focused too much on their professional lives, they'd likely never marry, and if they did marry and continued to work in high-powered jobs, the stress they would inevitably experience could make them infertile. Even some of those who were willing to pay that price believed that the premise was true. "Infertility," writer Susan Lang proclaimed, incorrectly, "was the unexpected fallout of the women's revolution."[11]

The medical profession itself was not immune to these attitudes. In 1982, the *New England Journal of Medicine* published a study of pregnancies after donor insemination, which concluded that fertility in women began to decline at thirty, not at thirty-five as previously believed.[12] The fact that the study found only a "slight decline" in fertility between thirty and thirty-five did not stop Alan DeCherney and his colleague Gertrud Berkowitz from advising young women in an accompanying editorial not to postpone childbearing to satisfy their professional ambitions. Consider having your children in your twenties, they wrote, and wait to start your career until your thirties.[13] Those at whom the advice was directed pushed back. Why, Harvard Medical School students Edith Brickman and John Beckworth responded, should women hold off on their professional dreams if they wanted to have children? A combination of "universal day care and equal parenting by father and mother" would be a much more appropriate way to address such a problem than the one proposed by DeCherney and Berkowitz. "Scientific facts," these students concluded, "in themselves do not lead to social conclusions."[14]

This editorial and the response to it also reflected a more general assumption within the medical profession that infertility rates were rising. In fact, the opposite was true. Infertility rates declined again, although only slightly, during the 1980s, from 8.5 percent in 1982 to 7.9 percent in 1988. And although the perception was that educated women had the highest risk for infertility, in reality the women most likely to suffer from the condition were African American women and women of all races of lower socioeconomic backgrounds. In 1982, when the infertility rate in white women was 8.1 percent, it was 13.1 percent in African American women.[15] And among whites, infertility rates were lower for college-educated women than those with less education. These data did not fit the dominant narrative, however, and the facts did little to change impressions. It was difficult to persuade physicians that infertility had declined in the 1980s, because greater numbers of women were seeking treatment. But there were more women in doctors' offices not because infertility rates were rising, but because the baby boom generation was larger than the generation before it by about twenty-one million people, and because this generation was more likely to seek medical treatment when they failed to conceive.[16]

Women who were in their twenties and thirties in the 1980s, having grown up in the era of the birth control pill, were primed to believe that reproductive decisions were under their control. Not only could they prevent unintentional conception, but if they used the Pill they were also accustomed to regular visits to a physician, because oral contraceptives were available by prescription only. Surely, with the Pill making it possible for them to choose *not* to get pregnant, they believed, when they stopped taking it, they would conceive. And if they failed to get pregnant on schedule, shouldn't there be a solution to that problem as well? Since many women already went to the doctor for contraception, at least they knew where to turn for diagnosis and treatment. The attention given to the first waves of what the press and the public invariably called "test-tube" babies in local newspapers and on television shows across the country put the issue of infertility in the news on a regular basis, making the idea of medical help for an infertility problem seem almost a given. Many of the stories also provided information on where to find that help. And if such advice was not readily at hand, would-be parents could turn to RESOLVE, which had become a national organization by the end of the 1980s.[17] Over the course of the decade, gynecologists and reproductive endocrinologists were seeing greater numbers of infertility patients; it was only natural for them to think that this increase in their patient population meant that infertility was on the rise.[18]

The Failure to Find Common Ground

Infertility epidemic. Man shortage. These were shorthand phrases for a larger effort to discourage women from believing that they—just like men—could successfully have careers, marriages, and babies. The idea that women should return to what was viewed as their traditional roles as wives and mothers was as much a part of the reproductive politics of the new Reagan-era conservatism as opposition to abortion. Both issues reflected intense political interest among Republicans in controlling women's bodies; linking them to the larger political context of the era allows us to understand their relationship to the new reproductive technologies.[19] Should these technologies be regulated? Should any of them be banned? The Democrats leaned toward regulation, but many Republicans could not support any form of regulation that allowed for the destruction of human embryos. This was the decade in which the Republicans became the anti-abortion party. Before the 1980s, many rank-and-file Republicans supported abortion rights. As late as the 1980 Republican presidential nominating convention, 51 percent of the delegates objected to a plank in the platform endorsing a constitutional amendment to ban abortion. But they were ignored, as were Republican feminists who objected to the removal of the plank endorsing the Equal Rights Amendment (ERA), which had been party policy since 1940. By the mid-1980s, feminist and pro-choice reproductive attitudes would no longer be part of mainstream Republicanism.[20] Reagan and the newly energized Republican right were tapping into a vein of cultural anxiety about the changing roles of women in American society, manifested both in Reagan's greater popularity among men than women and, by the time of his campaign for reelection in 1984, the overt emphasis he placed on his support for traditional family values and his opposition to abortion.[21]

The political signs pointed in too many directions to provide any sort of unambiguous guidance for policymakers seeking to develop some sort of national consensus on responding to the challenges of the new reproductive technologies. Pronatalist and anti-feminist attitudes are not necessarily antithetical to using technology to achieve a pregnancy. Anti-abortion conservatives of the 1980s, in contrast to the conservative ethicists of the 1970s who opposed IVF altogether on ethical and moral grounds, gradually accepted some reproductive technologies so long as no embryos were destroyed in their use. At the same time, opposition to IVF arose among some feminists who objected to what they called the "motherhood mandate," which they argued could drive women to employ risky and often unsuccessful procedures to become pregnant. The most intractable opposition to IVF came from a relatively small group of radicals called FINRRAGE (Feminist International Network of Resistance to Reproductive and Genetic Engineering). Renate Klein, one of its members, declared on the cover of her book, "Reproductive Technology Fails Women: It's a Con."[22] FINRRAGE insisted that IVF promoted the belief that a woman's only value rested on her ability to bear and rear children. When a *New York Times* reporter said to Janice Raymond, one of the founders of the group, that "women themselves are often beating down the door to the IVF clinic," Raymond responded that these women were being hoodwinked. "Recognize the political context," she retorted. "Women are submitting to pressure to have children at any cost because their lives are devalued without children."[23]

Saying that women who wanted to have children were dupes of the patriarchy denied that they had the ability and will to make their own reproductive decisions.[24] Women—including feminists—were actively involved in choosing whether and how to pursue infertility treatment and often were the primary decision makers in seeking medical help. It does not mean however, that they faced no pressure to reproduce in the first place. And while women's health activists overall were supportive of whatever reproductive decisions women chose to make, many of them were deeply concerned that a focus on the new technologies made pregnancy and childbirth possible for the few while taking attention away from such pressing matters as broad access to reproductive care, including abortion and maternal health.[25] In addition, the free market model in which IVF was offered made it difficult, and often impossible, for women of low and moderate income to afford it.

This swirl of conflicting voices came from interest groups with different and often competing priorities. Republican opposition to federal funding for in vitro fertilization research left them with little room for compromise, even as their Democratic colleagues made several efforts to find common ground. Between 1984 and 1987, four hearings were held in the House of Representatives on IVF and related matters. Two of them were aimed at the creation of a comprehensive national policy on the new reproductive technologies, and the other two were more narrowly focused. All four efforts failed. Not until 1988, when Ron Wyden, Democrat of Oregon, shifted the paradigm from regulation to consumer protection, did a legislative measure gain bipartisan traction. Wyden succeeded where others failed because his bill proposed the creation of a mechanism for clinics to provide information to potential consumers. It did not seek to make policy or regulate the development or use of these new technologies.[26]

The British Regulatory Model: Could it Work in the United States?

In 1984, Representative Al Gore of Tennessee, prompted by the progress being made in Great Britain toward regulating the new reproductive technologies, held hearings aimed at legislation that would create a similar model in the United States.[27] He had been closely following the deliberations of Britain's Warnock Commission, which had been appointed by the government of Prime Minister Margaret Thatcher and was chaired by Cambridge University philosopher Mary Warnock. It was charged with examining the new reproductive technologies and making recommendations on what "policies and safeguards should be applied, including consideration of the social, ethical, and legal implications of these developments." Its final recommendations included the creation of a "licensing authority. . . to regulate research and treatment" on them. The committee also supported "research on embryos . . . under license" for up to fourteen days after fertilization. Although it took six years for Parliament to legislate final approval of the commission's proposals, an intermediate "voluntary licensing authority" was created to regulate the technologies in the interim. The law, passed in 1990, was the result of a political process that allowed for competing interests to be heard, debates held, and compromises reached. The law also created the Human Fertilisation and Embryology Authority, known as HFEA, which has regulated assisted reproductive technologies ever since. If that kind of deliberative process could take place in Margaret Thatcher's England, Gore may have thought, why not here?[28]

Gore wanted to emulate the Warnock Commission's process as he sought to build consensus on the appropriate uses of the new reproductive technologies. "Other countries," he said, "have moved much faster than we have to consider where the appropriate limits are and what . . . values . . . are at risk." It was time for the United States to do the same.[29] The expert witnesses whose views he solicited were in general agreement on the need for an American version of the Warnock Commission, but not on much else. Biologist and ethicist Clifford Grobstein argued strongly in favor of a regulatory agency created by statute and with the power to rule on a range of issues, from defining the "status of the human embryo" to reducing "legal conflict and confusion," and maximizing "freedom of research while protecting essential human rights."[30] But the scientists and physicians who testified, including Howard Jones and Anne Colston Wentz, opposed outright regulation even as they endorsed the creation of an advisory council. They sought an end to the ban on federal funding, but they did not want Congress regulating how they chose to use the new reproductive technologies.[31]

The subcommittee members asked numerous questions about whether, and if so the extent to which, "human embryos should be afforded protection in law," which given the views of anti-abortionists came as no surprise.[32] With one exception, an official spokesman for the Catholic Church, all of the witnesses agreed that IVF using the sperm and eggs of a married couple was ethical; so was embryo freezing, as long as it was for later use by the genetic parents. Beyond that basic agreement, however, opinions diverged sharply. There was disagreement over the use of donor eggs and the sale or donation of "extra" frozen embryos to other couples. Several witnesses recoiled from the idea that couples might allow their potential genetic offspring—full siblings of their own children—to be gestated and reared by others. The most extreme views on the other side of that debate, at least in this period, were expressed by legal scholar and ethicist John Robertson of the University of Texas, who insisted that a married couple's "procreative liberty" should allow them to purchase any reproductive services they chose, including surrogacy, donor eggs, and donor embryos.[33]

It is hard to imagine Gore envisioning, as he listened to one expert argue against the use of donor eggs and embryos and a second declare that using them was a constitutional right, a middle ground on which everyone could meet. But he apparently did, declaring that the hearing had demonstrated the existence of "an emerging consensus that we urgently need a national commission on human genetic technologies [the mandate of which would include] the alternative reproductive technologies, so that we as a Nation can address these very difficult bioethical questions, and come to some sensible resolution of these questions." He went on to "urge his colleagues in the House and Senate to adopt that measure." He anticipated that his hearings would "get such a measure out of the conference committee."[34] It was a vain hope. There was little appetite among his colleagues in the Republican Party, which controlled the Senate, to deal with these questions, colliding, as they did, with the double wall of opposition to abortion and general antipathy to federal regulation.

The failure of legislation at this precise juncture, when the expansion of IVF into donor eggs and gestational surrogacy was in its early stages, had real and long-lasting consequences. As the decade wore on, with virtually no laws or congressional signposts to guide them, physicians became ever more reliant on their own moral compasses and institutional cultures to determine what uses of the new reproductive technologies were acceptable or beyond the pale. The American Fertility Society continued to issue guidelines, which many practitioners, but not everyone, tended to follow. But without a regulatory structure, it is no surprise that market forces continued to propel the development of the field, leaving legal questions to the province of individual states and the courts.

Is It Too Late? Democrats and the Regulatory Climate of 1987

After the Gore Hearings, it was three years before Congress tackled the subject again. The Select Committee on Children, Youth and Families, with Connecticut Democrat Bruce Morrison presiding, met in 1987 to examine the impact of the new reproductive technologies on children and the family.[35] In his opening statement, Morrison said that these technologies raised the issue of what "we as a society consider to be acceptable, to be equitable, to be legal, and to be sacred." The purpose of his hearings, according to Morrison, was to address such questions as whether American society "should focus resources on producing children through expensive technological methods when thousands of children await adoption, and when the children of low-income families suffer from inadequate prenatal care and nutrition." Should the nation be concerned about the "commercialization of human reproduction"? Does such commercialization "violate our most fundamental laws against trade in human beings," or is it "protected by Constitutional guarantees of the freedom to procreate?"[36]

At the heart of every one of these questions, Morrison said, was an attempt to understand two things: "How do we treat the fine line between reproductive choice for women and the risk of economic exploitation?" And "what role is appropriate or necessary for government on any level to take, as scientific discoveries outpace and potentially alter our social and legal framework?"[37] Morrison hoped that the findings of his committee would be used to shape public opinion and ultimately, national law. He felt encouraged that the subcommittee members present that day represented both political parties. Just as Gore had done three years earlier, Morrison called on professionals with scientific, medical, legal, or religious expertise, including some of the same people who had appeared in front of Gore's subcommittee. But times and attitudes had changed since Gore sought to create a commission to guide and regulate the development and use of the new technologies. Over the previous three years, with no federal rules in sight, new clinics had continued to open at a brisk clip, and pregnancies were being accomplished using frozen embryos, donor eggs, and now even "borrowed uteruses." If there had been a chance for prospective regulation, that prospect had faded. The technologies Gore had hoped to regulate in advance were already here. Morrison's reproductive landscape was different from Al Gore's, and several of his witnesses were unhappy with what they saw.

Wendy Chavkin, an obstetrician/gynecologist who directed New York City's Bureau of Maternity Services and Family Planning, decried the emphasis on in vitro fertilization and its offshoots, which she considered misplaced. Instead of seeking "a technologic fix to a problem with social roots," she said, Congress should concentrate on ways to improve the reproductive health of all women, including the poor, beginning with allocating "resources for basic health care needs." Chavkin was not opposed to the new technologies, she said, but she cautioned the legislators not to focus solely on care that only the affluent had access to, ignoring the needs of women of lesser means. Any technology available to the well-off should also be available to women of low income, she said. More importantly, the nation should focus its spending on reproductive and maternal health care for all women and on policies, such as paid family leave, that would allow women to "carry out both aspects of parenting and working simultaneously." If such policies were in place, she believed, fewer women "might find themselves at age 38 pursuing" a technological solution to their inability to conceive.[38]

The Democrats had not abandoned the idea of re-creating the Ethics Advisory Board, but Republicans remained opposed to the idea, with Dennis Hastert, Republican of Illinois, capturing their position when he objected to a national panel of experts becoming "the deciders of morality." If Republicans were concerned about embryo experimentation, the Democrats had their own moral qualms about the uses of the new technologies. Morrison was no more ready than Al Gore had been three years earlier to accept John Robertson's expansive concept of procreative liberty. Nor was he persuaded by the reasoning of another legal expert, Lori Andrews, who argued in favor of the enforcement of surrogacy contracts.[39] Both Democrats and Republicans were concerned about the expansion of the new technologies, but their attention was on different aspects of them. In an effort to bridge at least one of these divides—the question of whether states or the federal government had regulatory authority over them—Boston University ethicist George Annas proposed a possible compromise. States, he said, which regulate medical practice, should have primary jurisdiction over such technologies as "medical procedures . . . performed by physicians." But the federal government should also have a role, under its authority to regulate interstate commerce, in "forbidding the sale of human tissues, regulating false and deceptive advertising, and promulgating rules for human research."[40] Nothing came of his suggestions.

By the end of the hearings, Morrison felt profoundly discouraged. The witnesses were unable to agree even on the basic nature of the problem they were facing, which made it impossible for him and his fellow representatives to know where to begin. "What distresses me most," he said as diplomatically as he could before trying—and failing—to find common ground one more time in a final round of questions, "is that activities that are so well-developed out in the world already, are still the source of quite a bit of difference of opinion as to what our public policy approach ought to be."[41] The hearings ended on a note of futility.

Seeking Smaller Measures

Comprehensive regulatory legislation appeared to be ever more out of reach, it seemed, but perhaps, at least two legislators believed, it would be possible to address specific issues. A few months after the Morrison Hearings, in July of 1987, Patricia Schroeder, Democrat of Colorado, introduced a bill to require all insurance carriers that covered federal workers for obstetrical care also to cover infertility treatments, including IVF, and to pay a share of the cost of adoption. Schroeder called her bill the Federal Employee Family Building Act, and although its provisions applied only to federal employees, it was reasonable to hope that if the federal government were covering such treatments, other employers might follow. These hearings did not address such hot-button issues as embryo research or three-party reproduction but simply sought to lessen the financial burden on the men and women for whom the new technologies or adoption would allow them to fulfill their goal of having a family.[42] Schroeder's witness list was dominated by couples who had faced these problems and could testify to the ways in which infertility "undermines marriages, careers, and self-image."[43]

Schroeder wanted to provide some relief to infertile couples, an objective that comported with her larger legislative agenda, which revolved around support for women and families, a subject with both political and personal resonance for her. Provoked by a male colleague who questioned her ability to serve as an effective member of Congress because, after all, she was a mother, she famously retorted, "I have a brain and a uterus and I use both."[44] Schroeder was committed to the idea that government should do as much as it could to make it possible for every woman to use both those organs, too. In her time in Congress, Schroeder introduced legislation on pay equity, worked tirelessly to expand access to health care, and was the chief architect of the original, more generous, and unfortunately not enacted version of the Family and Medical Leave Act. She also proposed legislation regularly and continuously to promote research and expand services for contraception, infertility, and other aspects of women's health.[45] This bill was one of them.

During the hearing, witnesses testified to the significant economic hardship they faced when they found themselves needing IVF in order to conceive a child. One father told the committee, "We were very lucky, we could afford to try [in vitro fertilization] without insurance assistance." His wife's military benefits had covered an earlier round of medical and surgical infertility treatments, including tubal surgery. But the military did not cover IVF. The couple paid for it, he said, without having to go into debt. But "others we met in the waiting room told us about the problems they had in putting together the $4,000 for one try—people who had put together everything they could, borrowed from family, borrowed from friends, gone in hock to try the procedure once and didn't expect to be able to try it again." And they were luckier than the "many other couples" who were unable to get to that point, "couples who couldn't put together the money for even one IVF try," even if it was "their final hope."[46]

Schroeder placed the experiences of these infertile couples front and center, hoping that compassion for them and others like them might influence her colleagues. Her bill, too, failed to advance. About the only proposed legislation to gain even a modicum of traction was Democrat Thomas Luken's effort to criminalize the practice of paid surrogacy. Still, even with fifteen Republican co-sponsors, chief among them Republican Henry Hyde of Illinois, and four Democrats, this bill failed to generate much interest among his colleagues.[47] The Democrats were not interested in criminalizing surrogacy, and Republicans held to the belief that family law was the province of the states, not the federal government. If there was an interest in regulating or banning surrogacy, most Republicans believed, the states were the proper venue for such action.[48]

From Regulation to Consumer Protection

Al Gore had hoped that the example of Great Britain's Warnock Commission would encourage the United States to develop a similar regulatory process. Although it could be argued that the distinctions between our two political systems—federal here, parliamentary there—militated against such a development, a more important barrier to the consideration of adapting the British solution to American needs was the ideological chasm between the two political parties. Three years later, the ideological divides had grown even wider, and new ones had opened. Representative Morrison found, to his dismay, that disagreements were so fundamental that he could find little consensus about the nature of the problem, let alone its solution. In the end, neither Gore nor Morrison was able to find a formula that would lead to agreement on the regulation of the new technologies.

One of the principal casualties of this stalemate was federal funding for human embryo research. Leading researchers and practitioners in the field of reproductive medicine had continued to urge Congress to support such funding. However, these physicians were also becoming more likely to oppose federal regulation of the new reproductive technologies as interference with their professional autonomy. Add to this mix the staunch opposition to such research from the organized anti-abortion movement, which included the Catholic Church and many evangelical denominations, and it was clear that this problem could not be resolved in the existing political climate. Beyond these divisions, there was disagreement over questions of regulation more generally. Democrats looked to federal authority to oversee these new ways to reproduce. Republicans looked to the states, not the federal government, to address legal issues pertaining to family life. In short, by the late 1980s, there was no American consensus on the regulation of in vitro fertilization at a time when the technology was expanding both geographically and in terms of the kinds of procedures offered.

By the end of the decade, in vitro fertilization was available almost nationwide, and a few states had begun to address the problem of affordability, requiring employers to provide insurance for it. In 1985, Maryland became the first state to mandate insurers to cover IVF. By 1987, it had been joined by three other states, Arkansas, Hawaii, and Massachusetts, although the coverage was somewhat restrictive in the first two. Texas mandated that insurers offer coverage for IVF, but employers were not required to purchase that coverage.[49] (See table A.1.) Everywhere else, patients were not only on their own in terms of paying for IVF, but they were also on their own when deciding how to select a clinic or physician to provide treatment.

This was a moment made for Ron Wyden, then a member of the House of Representatives from Oregon, who shifted the legislative conversation about IVF from regulation to consumer protection. Wyden had come to believe that by framing this issue as one of providing essential information to enable patients to make rational choices about providers and costs, he could strike a compromise among the multiple competing interests of physicians, patients, legislators, and the public, thereby creating some level of oversight over the procedure. Wyden did his homework. He and his staff reached out to physicians, sending out a questionnaire to the nation's IVF programs that yielded a remarkable 87 percent response rate. He also held two sets of hearings in 1988 and 1989.

The responses to the questionnaires, which revealed how rapidly in vitro fertilization was spreading across the country, helped to make a case for Wyden's approach. In 1987, about 8,900 women across the United States, in centers large and small, had attempted in vitro fertilization at least once. In 1988, that number was up to 11,500.[50] These numbers, Wyden said, were significant. In his opening statement, he described the new field of assisted reproduction as a vast, unregulated commercial enterprise. "An estimated $1 billion," he said, "will be spent [this year] on treatments in the United States."[51] But no one, he argued, was monitoring the providers of such treatments. The American Fertility Society (now the American Society for Reproductive Medicine) provided voluntary guidelines, but that was all. As Wyden accurately pointed out, "There are no Federal or State requirements for accreditation or reporting. There is, in fact, no public accountability at all."[52]

Wyden was aware of the failed efforts of his colleagues who had attempted to create an overarching national policy or develop regulatory requirements. He chose a different path, tapping into a powerful consumerist movement among American patients who increasingly viewed their relationships with doctors in transactional terms. Here was an area of potential agreement between the two parties. Who could argue against providing information to enable patients to become better consumers of medical services? As historian Nancy Tomes has noted of this phenomenon more broadly, "A pattern thus emerged in the late 1980s that would repeat many times in the years to come: appealing to new standards of public accountability, patient-consumers would gradually receive access to data formerly thought suitable for experts to see and interpret."[53] Wyden, in short, made it his mission to protect couples in their role as *consumers* of reproductive services. The shame of the new technologies, in his opinion, was in the money spent with no understanding of the chances of success. The existing situation was "a prescription for disaster," he said, with couples, "many of them desperate, spending huge sums of money on technology that has been sold through borderline advertising." Or, as *Time* interpreted Wyden's remarks, "If a husband and wife put down $7,000, they have a right to know what chance they have of getting a joyous return on their investment."[54]

Although the Wyden bill was specifically designed with IVF in mind, and not the larger field of reproductive medicine, his investigations were inspired by the shocking allegations made against Cecil Jacobson, a prominent local infertility specialist who did not even offer IVF in his medical practice. Jacobson did, however, defraud his patients. Vicki and Bill Eckhardt testified that between their own money and that of their insurance company, they were fleeced out of nearly $35,000. Jacobson had repeatedly injected Vicki with what he said were fertility drugs, told her she was pregnant, and then claimed that her fetus had died and been absorbed into her uterus. He was so plausible that the couple believed him through seven so-called pregnancies. After that, their suspicions awakened, they saw another doctor and then spoke to other patients who had also been defrauded. In addition to his pregnancy scam, Jacobson cheated other patients in a different way, using his own sperm for artificial inseminations in his medical practice but claiming that the sperm had come from anonymous donors. Not until his perfidy was revealed in these hearings was his medical license revoked. He was arrested, convicted on fifty-two counts of fraud and perjury, and sentenced to five years in prison.[55]

Shockingly, until his long-standing unethical and illegal activities came to light, Jacobson had been a well-respected doctor. If he could get away with this kind of behavior for so long, Wyden wondered, how many others like him were out there? And now, with the rapid growth of the new reproductive technologies, how could patients find a fertility specialist they could trust? Could they believe the advertisements of a fertility clinic? Maybe. Maybe not. And what about the embryology laboratories associated with those clinics, in which women's eggs were fertilized? Some were good, others perhaps not. Given that there were no certification requirements, how could anyone tell? Patients had no way to know what they were getting, Wyden said. At the very least, he believed, there had to be a way for them to learn the success rates of the IVF clinics they were considering and to have confidence in the skill of their embryologists.

This, it turned out, was the magic formula for legislation: Wyden crafted his legislation as a consumer protection measure at the very moment that the idea of providing information to medical consumers was extremely popular across the political spectrum. He was extremely savvy in promoting his bill, seeking and receiving support from specialists in reproductive medicine in the American Fertility Society and its affiliate, the Society for Assisted Reproductive Technology, and enlisting the patient advocacy community, represented by RESOLVE, which saw consumer protection laws as essential to keep patients out of the hands of charlatans.

By the time of the Wyden Hearings, physicians active in the development and use of IVF had soured on the idea of outright federal regulation, convinced that professional guidelines developed by the American Fertility Society would ensure compliance with ethical standards. Wyden acceded to this majority view, including in the bill a provision prohibiting either the secretary of health and human services or state legislatures from establishing "any regulation, standard, or requirement which has the effect of exercising supervision or control over the practice of medicine in assisted reproductive technology programs."[56] As introduced in 1990, the Fertility Clinic Success Rate and Certification Act, known informally as the Wyden Act, required clinics and programs who treated patients with what would come to be called assisted reproductive technologies—that is, all procedures in which eggs or embryos are handled outside the body—to report their success rates annually to the secretary of health and human services through the Centers for Disease Control and Prevention. The law was passed by voice vote in both houses of Congress and signed by President George H. W. Bush in 1992.

Wyden succeeded where others failed because his legislation deliberately did not address larger policy issues. Its language rejected regulation outright. Instead, it created a mechanism by which consumers could view the success rates of a clinic and compare its rates to others.[57] Compliance from the clinics was mandated, but because the law had no enforcement mechanisms or legal penalties for failure to report, it created what was essentially a voluntary system, relying on the willingness of doctors to submit their information. The Wyden Act provided consumers with the tools to examine clinic success rates, but it constituted neither policy nor regulation. It did not address the important issue of funding for research in the field. It did not regulate the assisted reproductive technologies.[58] There was significant peer pressure on physicians to report their success rates, and those who failed to do so were denied membership in the Society for Assisted Reproductive Technology. Nevertheless, doctors who refused to comply with this law, unless they were violating other state or federal laws or engaging in actions that would result in the loss of their medical license, could continue offering these technologies to patients. As it turned out, however, urging doctors to report their results to a central database was the best that could be hoped for in America's fractious political climate.[59]

The Deeper Meaning of Regulatory Impasse

Assisted reproductive technology developed in an era when American culture was deeply divided over questions of reproduction, sexuality, and the meaning of family life. Even the idea of family had become hotly contested. On the one hand, many religious conservatives wanted Americans to re-create what they saw as the traditional family form of two parents, one of either sex, with a breadwinner father and a homemaker mother. On the other hand, women's rights advocates and others argued that such a goal was both unrealistic and undesirable. Women's roles had changed, they said, and even some non-feminists could see that most families required two earners to hold their ground in the middle class. But that did not quell the anti-feminist backlash. By the late 1980s, it was in the ascendancy, with high-achieving women being urged to abandon the fast track for the mommy track.

Over the course of the 1980s, some religious conservatives accepted the new technologies, at least in part. So long as no human embryo research was allowed, and no embryos were destroyed, in vitro fertilization using a husband's sperm and his wife's eggs could be viewed, as it was during the John Rock era, as simply an unconventional way to create a conventional family. On the other side of the political divide, there were some feminists who objected to reproductive technology because, they said, it pushed the idea of a "motherhood mandate" that undermined women's autonomy. Women's health activists urged the nation to pay attention to women's health needs more broadly, to protect access to abortion, and to be less concerned with providing offspring for the few. Feminists also opposed the free-market model for access to the assisted reproductive technologies, which meant that women and couples with low or moderate incomes could not afford it.

The politics of reproduction had become so complicated and messy during the 1980s that compromise and consensus grew ever more elusive. Only by reframing federal oversight of assisted reproductive technology as "consumer protection," outsourcing the collection of data to the professional organization for reproductive endocrinologists, and eliminating any hint of outright federal regulation could the federal government pass legislation to address the provision of these new technologies. By then, the numbers of clinics offering IVF had grown dramatically to nearly two hundred by the decade's end, and an estimated thirty thousand women in the United States had sought a pregnancy using IVF.[60]

Lack of regulation meant that physicians were left to make their own rules. By the end of the 1980s, embryo cryopreservation (freezing) had become more common. Many clinics continued to limit their practices to married heterosexual couples up to the age of forty using their own sperm and eggs, but others were expanding those age limits into the forties. The field itself was changing. Over the course of the next decade, both the supply of doctors trained to provide IVF and the demand for their services increased. There was more competition for patients and more interest in expanding the patient base beyond the traditional infertile couples using their own sperm and eggs. Among the potential new markets for the technology were heterosexual couples interested in donor eggs or embryos or gestational surrogacy, same-sex couples, single women, and postmenopausal women. Without the prospect—or, as some saw it, the threat—of regulation, commercialization became the dominant theme of the next decade.

6

A LOT OF MONEY
BEING MADE

In the summer of 1995, Ricardo Asch, a prominent infertility specialist at the University of California, Irvine, told his colleagues that he was leaving town for an international lecture tour. He wasn't. He was fleeing the United States to avoid arrest after federal prosecutors had raided his home and office. The previous spring, the *Orange County Register* claimed that Asch had "harvested eggs from an Orange County woman without her consent and gave them to another patient who delivered a baby boy about nine months later." This accusation was only the first of many. "If those allegations hold up," bioethicist Arthur Caplan told the *Register*, "they would be the most serious violation of ethical trust that I am aware of in the field of reproductive technology." Added Caplan, "There may be worse things that one could do in operating a fertility clinic, but I don't know what they are."[1] By the following November, the paper put the number of women who had unknowingly had their eggs and embryos taken and used in other women at sixty and climbing.[2] The allegations against Asch and his two medical partners made headlines across the country, brought heartache to scores of couples, sent shock waves through the profession, and cost the university more than $27 million in legal settlements to affected patients.[3]

In the 1980s, the Argentine-born Asch had been a hotshot young reproductive endocrinologist at the University of Texas Health Science Center in San Antonio. He had arrived in the United States in 1975 for a postdoctoral position in endocrinology at the Medical College of Georgia, where he worked under Robert Greenblatt, a major figure in the field of reproductive medicine who was best known for his discovery in 1961 that clomiphene citrate could be used to induce ovulation. Asch left Georgia for San Antonio in 1979, where he completed a fellowship in reproductive endocrinology, staying on as head of the Division of Human Reproduction in the Department of Obstetrics and Gynecology. San Antonio's attempts to achieve an IVF birth were not successful, but in 1984, Asch and his colleague Jose Balmaceda achieved the world's first birth using gamete intrafallopian transfer, known as GIFT.[4] The development of GIFT made Asch a bona fide reproductive medicine celebrity at the age of thirty-seven. Two years later, he was recruited to be director of the AMI Center for Reproductive Health Care of the University of California, Irvine.[5] Balmaceda joined him. Asch's rise occurred as the new reproductive technologies were beginning to provide lucrative business opportunities for those inclined to seek them out. He took advantage of those opportunities and by the early 1990s achieved the kind of wealth that most of the early pioneers in the field had never even thought about, let alone sought.

Like Asch, some doctors were embracing the business of IVF, just as many infertile couples thought of themselves not only as patients seeking treatment for a life-altering condition but also as consumers of a medical product. This idea had received legislative legitimacy with the Wyden Act, which shifted the governmental role from failed attempts at regulation to consumer protection. At one level, the change was simple recognition of what laws could accomplish given the fractured political landscape. But at another, it reflected the reality of IVF, a costly procedure for which most people paid out of their own pockets. IVF programs, known as clinics at the time the Wyden Act was being written, were becoming fertility centers, and many of them were privately owned and independent. They competed for market share. Patients became consumers of a luxury good. Even as we recoil from calling a baby a "product," the idea had found its way into public discourse.

If patients who were paying fifteen, twenty, or thirty thousand dollars in their quest for a baby thought about their transactions with physicians at least in part in business terms, who could blame them? If Ron Wyden came to believe that consumer protection legislation offered his best or perhaps his only hope of getting some control over this unregulated field, who could blame him? If physicians sought to prove to potential patients that their IVF clinic was better than the one across town, who could blame them? If some of those physicians resorted to unethical practices in order to boost their success rates, not to mention their income, we *can* most surely blame them. But should we be surprised?

When the medical school at UC Irvine recruited Asch to its faculty, it put a high value on his reputation to bring a major new stream of income to the university. It was not as if the medical school lacked talent in this field. It already had Bill Yee, who directed his own, separate, university-affiliated IVF clinic at Memorial Hospital. Nevertheless, the university seemed almost literally to be banking on its new young superstar from Texas. "The fact that Asch appeared to be a medical money-making machine was perhaps [his] most appealing attraction" to UC Irvine, two scholars who studied the scandal wrote a few years later.[6] Asch delivered on the medical school's financial expectations. His practice revenue was said to be in the multimillions. The more money Asch and his clinic brought in, the more money the university made, since a percentage of his clinic's revenue accrued to the medical school. That might indicate, some said, why the university seemed slow to act on complaints of irregularities in the clinic made by members of Asch's staff. When UC Irvine did investigate, its auditors discovered, although the claim was never independently verified, that Asch had withheld more than $7 million of his clinic's income owed by contract to the university.[7]

According to *Newsweek*, Asch enjoyed a lavish lifestyle, keeping "thoroughbred horses, driv[ing] a Ferrari and own[ing] a couple of million-dollar homes." He and his family lived in the wealthy seaside community of Newport Beach, their house decorated with original works by Salvador Dali. Their vacation home was a $2 million beachfront house in Del Mar, outside of San Diego. Asch hobnobbed with celebrities, was a habitué of the racetrack, and owned five racehorses. A bit unnervingly, given the deeds of which he was accused, two of those horses were named "Fibs Galore" and "Golden Find."[8]

We are not suggesting that making money, even making a great deal of money, is or should be a crime. But fraud is a crime, and medical negligence can be. The accusations that Asch's practice had taken eggs and embryos from patients, using them for fertilization and implantation in other women without consent from any of the patients involved, riveted the state and the nation. If the reports were true, one of the leading figures in reproductive medicine, a faculty member at a major medical school that was part of one of the most distinguished systems of higher education in the country, had engaged with impunity in serious medical misconduct for several years. Asch and his partners, Jose Balmaceda and Sergio Stone, consistently denied these shocking allegations. Stone was ultimately tried and convicted of fraudulent billing; Asch and Balmaceda fled the United States and never returned. Asch successfully practiced in Mexico City and consulted widely in Latin America and elsewhere. By the time he was arrested in Mexico in 2010, he had been a fugitive for fifteen years. But Mexico refused to extradite Asch, effectively ending the attempts of the United States to prosecute him. Balmaceda, who had been born and educated in Chile, returned there to practice medicine. Stone was convicted, paid a fine, and had his medical license temporarily suspended. His reputation was in shreds.[9] The scholars who subsequently examined the scandal in detail put the onus for the misdeeds on Asch. Stone, they concluded, had not been at all involved in the misuse of eggs and embryos. Balmaceda knew that the practice was occurring and failed to report it, they said, but he did not engage in it himself.[10]

We may never know the entire story. We do know that eggs and embryos were misappropriated. Alan DeCherney, one of the experts dispatched to Irvine to examine the clinic's records from the National Institutes of Health, told us in 2015 how easily it could be done.[11] Asch, he said, might get "six eggs from one person and no eggs from another one. He told them that he got three eggs from each. Both women were inseminated with their own husband's sperm. Both people had embryos put back."[12] Except for a small group of women who said they had consented to donating some of their eggs, the patients whose eggs and embryos were taken had no knowledge of what had happened. Neither did the couples who received them.

DeCherney also recalled that his initial impulse was to take Asch at his word when he denied all the accusations. After spending an "afternoon counting transferred embryos and eggs," he told us, "I didn't find anything wrong. Everything added up, 100 percent. I went into the room where the people from NIH were, and I told them that I couldn't find anything. 'There's nothing wrong, here,' I said. 'Everything added up.' There was one guy at the end of the room, and he said to me, 'Doctor, do you have a checkbook?' I said, 'Yes, we have a checkbook. My wife used to keep the checkbook. But she could be off a couple of dollars, and that didn't make sense to me. After all, she had a calculator. So I took it over.' And the NIH guy asked, 'How did you do?'" DeCherney replied that his check-balancing skills were much worse than his wife's. Asked how many entries he made to do so badly, DeCherney said, "'20–25 entries.' And then he said, 'So, here are these people making thousands of entries and they are not making a single mistake. Aren't you suspicious of that?' I said, 'No. Maybe they're careful.' He said, 'They are cooking the books.'" In the end, DeCherney said, the "NIH guy" turned out to be right. "They did cook their books, and it was discovered."[13]

If the allegations are correct, what could have made Asch treat his patients' genetic material so cavalierly? Hubris? Ambition? Perhaps a lurking worry about whether he could live up to his early reputation? Asch had shot to prominence in 1984 with GIFT, and initially many of his peers thought that for some conditions GIFT would be just as important as IVF. When UC Irvine recruited Asch, he had seemed to be doing everything right—developing a new technology, publishing in all the right journals, and getting his patients pregnant.[14] But by the end of the decade, as IVF techniques became less invasive and IVF success rates for all diagnoses rose, the use of GIFT had declined dramatically, and by the mid-1990s, it was used only infrequently.[15] Asch was an IVF specialist, too, but GIFT had made his reputation. Was he now becoming worried about keeping his IVF success rates up to hold on to his place in the highly competitive California environment? Did he believe it was his job to help a couple to have a child even if he used someone else's genetic material without anyone's knowledge? Could he have done it just for the money? Alan DeCherney told us he still finds it hard to believe that Asch was driven by financial motives. DeCherney was not defending anyone's indefensible actions; he was simply trying to understand what could motivate any doctor to breach his patients' trust in this way.[16]

A few months after the scandal broke, Asch told journalist Diane Sawyer that one of his five children, his only son, was adopted. Environment, not genetic makeup, he said, was the critical factor in his son's—in any child's—development. "Genes are, in my opinion, not that important," he said. "I think it's entirely obsessed, this society, with genes."[17] Some heard those statements as an expression of indifference to the feelings of patients who felt "deprived of a biological child."[18] Regardless of Asch's motives, such hubris is astonishing. Who was he to decide on the importance of genetic parenthood for anyone but himself? Perhaps genes meant nothing to him, but they did to his patients. The couples whose genes, embodied in their eggs and embryos, were stolen from them suffered deeply from the knowledge that those eggs or embryos might have produced children whom they would never see or know. At least fifteen children were born from those misappropriated eggs and embryos. John and Debbie Challender were devastated to discover that their embryos were given to another couple. In their case, they knew that the embryos had resulted in the birth of twins. John told the *Register* that he could not stop thinking about those children. "The embryos they took were our children," he said. "The embryos they stole, those were my children."[19]

Asch had his defenders among patients whom he successfully treated and who could not believe he engaged in unethical practices.[20] Asch himself insisted that if there *were* any misappropriated eggs and embryos, it was the staff's fault, not his. According to the *Los Angeles Times*, when Asch met with US attorneys at a hotel in Mexico in 1996, he said under oath that he "was not involved in running the clinics beyond performing surgery" and did not personally "match donors and recipients, did not obtain patient consents and had no way of knowing how any mistakes occurred."[21] Stone and Balmaceda said they knew even less. In the end, all three were indicted for fraud and misreporting income.[22] Meanwhile, the University of California was sued by numerous couples claiming that each egg or embryo taken from them meant the loss of a potential child.[23]

Asch's behavior and the misdeeds he was said to engage in were on such a grand scale that one might be tempted to dismiss the entire episode as a sensational story involving one IVF clinic at a single university. Or was it, as Balmaceda told researchers studying the scandal, simply business as usual? "Much of what went on at the university fertility clinic, whether it was unconsented egg appropriation or illegal insurance billing," he claimed, "represented the normal course of business in the United States."[24] Balmaceda provided nothing else to back up that assertion, however, and we have not been able to corroborate any such claim, either. It was true that there were some clinics that lied about their success rates and were sanctioned by the Federal Trade Commission for doing so.[25] They were not, however, stealing eggs and embryos. Nevertheless, if Asch's alleged misdeeds were not typical, his behavior should have served as a warning of the kinds of things that could go wrong without checks on a doctor's behavior. Asch nominally worked for a university, but there was virtually no oversight of his fertility center. He was essentially left to run a lucrative private enterprise however he saw fit. The Asch scandal provided a focus for what was otherwise a growing societal unease about the directions in which the new reproductive technologies were moving in the 1990s. It also laid bare the lurking question of trust. How could patients know whom to trust in a field where treatment possibilities were becoming more complex and more expensive, and that remained largely uncontrolled? The Asch story brought home to many couples how little protection they had.

Alan DeCherney had already been concerned in the 1980s about "the potential for the conversion of medicine, at least philosophically, into a business."[26] Ten years later, that potential had become a reality. Many of his colleagues had come to embrace the business model. Even in academic medicine there were directors of IVF programs who felt the lure of commerce, some of them developing formal affiliations with for-profit companies. When Mark Sauer became the director of the division of reproductive endocrinology at Columbia University's medical school in the late 1990s, he partnered with a company called GynCor, which invested $1.5 million in Columbia's facilities.[27] It may have been that experience that led to Sauer's unease about where such enterprises were leading the field. Assisted reproductive technology, he said in an interview in 1999, had become "a business, as opposed to the academic pursuit of a university." One attraction of the private sector for some, he added, was that "there was a lot of money being made." Privatization was "inevitable," he continued, but he expressed concern about how "commercialization" was affecting those practicing in the field.[28]

Sauer was right. In the 1990s, large private fertility centers took an increasing share of the market for assisted reproductive technology. Of the ten largest IVF clinics in 2000, only three were housed at medical schools in academic departments.[29] IVF programs at academic medical centers remained important for many reasons, but in terms of numbers, the for-profit centers were becoming more prevalent. In addition, assisted reproductive technologies were attracting new types of patients. Heterosexual couples of conventional reproductive age using their own sperm and eggs continued to make up the overwhelming majority of IVF patients nationwide, and still do, but they were no longer the only market. In the 1990s, paid egg donation, gestational surrogates, and new technologies and procedures expanded the technology's possibilities beyond simply fertilizing a woman's egg and then transferring it into her uterus. Gaining considerable attention, they also raised new and complicated questions. How should society deal with issues of access when the well-to-do could easily afford any of these new technologies, but low- and middle-income couples and individuals were either shut out or forced to go deeply into debt simply to have a child? How should motherhood be defined when a surrogate conceived, carried, and gave birth to children to whom she was not genetically related? Who is the "real" mother of a child conceived with another woman's egg? Would same-sex couples have the same ability to become parents as heterosexual couples, and if so, what role in assuring them their rights would be played by assisted reproductive technology?

In the final decade of the twentieth century, Americans struggled with these and other questions in a reproductive environment with few rules or guideposts. When IVF began, most reproductive endocrinologists had thought of it as a new option for infertile women in their twenties or early thirties with tubal disease. Now, however, the technology seemed to have taken on a life of its own. The Wyden Act had reified the notion of doctors as purveyors and patients as purchasers, seeking to protect the latter in that role; and the only other national policy was the ongoing ban on federal funding of all research in which an embryo might be destroyed. Patients could look up clinic success rates, and abortion opponents had effectively stymied independent, federally funded research. In certain states, lawmakers and judges had attempted to address issues such as whose wishes should prevail in cases where one partner wanted to use a frozen embryo and the other wanted it destroyed, or whether gestational surrogacy contracts should be honored or declared unenforceable, creating a confused and conflicting legal tangle for potential parents to navigate.

Growth and Expansion in the 1990s

In 1989, around 13,500 women were treated with IVF at the 163 clinics that had voluntarily reported their success rates to the Society for Assisted Reproductive Technology. Two thousand of those women, just under 16 percent, gave birth, and 95 percent of those births were to married women using their own eggs and their husband's sperm. The most common condition that brought these couples to the IVF clinic was tubal disease, followed by endometriosis and a condition called unexplained infertility.[30] All in all, between 1981 and 1989, around 5,000 women in the United States had babies whose births had been made possible by IVF.[31]

Flash forward ten years, and in 1999, there were 370 programs offering IVF and related technologies, located in almost every part of the United States, and that year alone, almost 22,000 women gave birth after IVF. Success rates, measured by the birth of a baby, had risen to almost 30 percent per egg retrieval. The use of donor eggs had risen, and nearly 14 percent of births were to women who had used them. Some of the rise in donor egg use could be attributed to the greater prevalence of compensated egg donation. Gestational surrogacy was rare, accounting for just 1 percent of births. During the 1990s, almost 81,000 women had babies using assisted reproductive technology.[32]

There were other changes as well, besides in the numbers of patients seeking care. It was still true that most IVF patients were younger than forty, but they were slightly older than their counterparts in the 1980s. (See figure A.1.) The primary reason for this shift was that more couples were marrying and starting their families at older ages. By 2000, the rate of first births to women between the ages of thirty-five and thirty-nine had risen to 8.9 out of every thousand births, up from 6.7 just a decade earlier. Births to women between forty and forty-four also had risen. In 1983, the rate of first births to women between forty and forty-four was 0.4. In 2000, it was 1.7.[33] Women over age thirty-five having their first baby was big news in the 1990s. Stories regularly popped up in the media about women who "nonchalantly put off motherhood" until their late thirties, believing that it would be as simple to become pregnant as it had been to avoid pregnancy, only to discover that it might be too late. The quest for pregnancy, journalists reported, "ruled the lives" of these surprised women. "If you don't have a child, and you want one," said one mother who had a baby after IVF, "you'll take every possible chance . . . It's the hope that drives you on."[34] With varying degrees of urgency, these women and their male partners sought medical help. Sometimes the pull of having a genetic relationship, sometimes the desire to experience pregnancy, motivated them. They did not want to be denied what others seemed to achieve so effortlessly.[35]

In 1995, about 11 percent of women between the ages of twenty-five and forty-four had visited a doctor at least once for medical help to become pregnant. Just a tiny fraction—one-tenth of 1 percent—used assisted reproductive technology, most commonly IVF. Women with college degrees and higher incomes were more likely to do so.[36] African American women as well as women with less education and lower incomes had fewer choices overall when it came to dealing with infertility. Access to medical treatments and assisted reproductive technology was stratified by both class and race.[37] During the 1990s, most IVF patients were white. African Americans appear to have accounted for just 4 percent of all the IVF cycles initiated. As sociologist Dorothy Roberts straightforwardly put it, "The people in the United States most likely to be infertile are poor, Black, and poorly educated. Most couples who use IVF and other high-tech procedures are white, highly educated, and affluent."[38]

In fact, they were a lot like M-Liz Reicher and her husband, who had spent four years trying to have a baby by the time she turned thirty-five. She was a communications specialist at a consulting firm and her husband, Gene, was a vice president at a media company; they lived in the suburbs of Washington, DC. M-Liz was thirty when they married, and they were eager to become parents. Luckily, they had the financial resources to move on to IVF when medical treatments failed. After four attempts, she conceived. They were delighted with the birth of their daughter Megan and had no regrets about spending the money. Still, it was an extremely expensive process. All in all, they said, the total cost of their treatments had amounted to approximately $100,000.[39] Parenthood was equally important to couples who lacked the resources of the Reichers, but they often had fewer options. Some of them wiped out their life savings, borrowed from family members, or took out second mortgages to pay for treatment. A cycle of basic IVF might cost $7,000 (about $12,200 in 2018 dollars) or more. One Maryland teacher, Bob Lenz, said that "without insurance," he and his wife, who underwent four IVF cycles before becoming pregnant with their son, would have been hard pressed to cover the cost of treatment. A Pennsylvania father, a well-to-do business owner whose state, like all but a few, did not mandate insurance coverage for IVF, said, "I could afford it. But why," he asked, "should anybody be denied" what he was fortunate enough to be able to afford?[40]

Another couple, the Lenzes, lived in Maryland, where IVF was covered under employer-sponsored health insurance plans in the 1990s, which enabled couples with moderate incomes to take advantage of the technology. By 2000, thirteen states required health insurance plans either to cover or to offer coverage for some kind of infertility treatment, but the laws were wildly incongruent. (See table A.1.) For example, California's law excluded IVF but mandated coverage for GIFT, which by the turn of the twenty-first century was so rarely used that it had become virtually obsolete. Some states, including Texas, required plans offered by health maintenance organizations (HMOs), but no other insurers, to offer IVF coverage. To make it more complicated, even though the HMOs in Texas had to offer it, employers did not have to purchase it. Some states required a couple to exhaust less costly forms of treatment—even if those were failing—before paying for IVF.

If these complexities leave the reader reeling, just imagine how difficult it was for patients. At the turn of the century, only five states—Connecticut, Illinois, Maryland, Massachusetts, and Rhode Island—mandated more or less comprehensive coverage for infertility treatment, including at least a limited number of cycles of IVF. But even in the states mandating coverage, an IVF program could refuse to accept insurance and instead charge a set fee, making the patient pay the difference between what the insurance covered and the clinic's fee for the procedures.[41] IVF clinics set their own charges. Doctors could lose business if they charged significantly more than their competitors, but there was no other limit on cost.

Donor Eggs, Host Uteruses, and Frozen Embryos

Richard Marrs remembered that his early donor egg patients were women in their late twenties who suffered from premature menopause. Most of them brought along their own donors—sisters, cousins, close friends. The use of donor eggs was not widespread during the 1980s, and where it was used, most of the patients were like those of Marrs. The Jones Institute's donor egg program in the 1980s used two kinds of altruistic donors. Some were providing eggs to a family member or friend, and others were women undergoing IVF themselves, who agreed to share their eggs anonymously with women unable to conceive with their own.[42] But not everyone who required a donor egg to conceive had a sister, best friend, or generous stranger willing to part with some of hers, which meant that if IVF clinics chose to use donor eggs, their use would have to be regularized. From sisters to brokers who were themselves paid to purchase an egg from a stranger—that was the trend. Marrs recalls that as the number of patients wanting to use donor eggs expanded, and with fewer coming in with their own donors, he decided to recruit donors for them. One of his staff members, Mirna Navas, agreed to take on the task. Navas ultimately left his employ to start her own donor egg agency, which Marrs still uses. Many fertility centers use such agencies, while others continue to recruit their own donors.[43]

As the number of clinics providing donor egg services increased, the process became more commercialized. In the western United States, brokerages became more prevalent, especially in California. On the East Coast, many IVF centers maintained their own donor pool. In either arrangement, payment for eggs were the norm, and the number of known donors dropped. In 1993, 22 percent of donors were known to the recipients. The following year, it was down to 17 percent, and after that the proportion was apparently so small that the statistics were no longer tracked.[44]

Although donors and recipients were now strangers, there was a desire on the part of both to have some sense of personal connection. Even in programs where mutual anonymity was the rule, recipients received information about the donors, including their medical histories. Sociologist Rene Almeling, an expert on egg and sperm donation, writes that anonymity has remained the common practice in the eastern United States, but on the West Coast, recipients can meet and select their own donors.[45] In these early years, compensation for egg donors was below $1,000, but the procedure was expensive nonetheless, because of the cost of a range of medical services for donor and recipient.

Until the mid-1990s, most clinics that offered donor egg services did not treat women over the age of forty-two or forty-three. In the 1990s, however, studies conducted by Mark Sauer, Richard Paulson, and Rogerio Lobo demonstrated that that women in their forties and even fifties could become pregnant with donor eggs and carry the pregnancy to term. The three doctors made the case that women in these age groups should not be denied this service, insisting that it was a matter of "reproductive choice."[46] As Sauer told reporter Kathleen Doheny, he "steadfastly" believed "that almost any woman who desperately wants to be a mother should have the chance, whether she is young or old, married or single, heterosexual or gay." Even so, he refused to treat women over the age of fifty-five, because it took him out of what he called "his personal comfort zone."[47]

It was only a matter of time before that age limit was breached. In 1996, a woman claiming to be fifty years old appeared in the office of Sauer's fellow researcher Richard Paulson. He accepted her as a patient, she conceived using a donor egg, and gave birth. She was not fifty, it turned out, but sixty-three, his oldest patient by far. Still, even in this practice, where women up to their mid-fifties were welcomed, the average age of donor egg recipients was forty-three.[48] Elsewhere, that average was closer to forty or forty-one. In the 1990s, few women over age fifty sought the service; even women in their forties were a smaller share of the donor egg market than those in their thirties. But now that doctors knew it was possible for women to conceive and bear children well into their fifties, the temptation was surely there to offer the service to women who requested it.[49]

The use of donor eggs increased during the 1990s. Gestational surrogacy, however, remained rare, accounting for just about 1 percent of all births after IVF at the end of the 1990s. Of all the uses to which assisted reproductive technologies were put, surrogacy was among the most controversial, receiving an outsized share of public attention. Even when it was done for the most altruistic of reasons, many Americans were disquieted by the idea that a woman would intentionally conceive and bear a child whom she did not intend to raise. When Arlette Schweitzer, a forty-two-year-old librarian from South Dakota, became a surrogate for her daughter in 1991, she faced considerable criticism. Eight years earlier, when Christa was fourteen, Arlette had taken her to the Mayo Clinic because she had not yet begun to menstruate. There she was diagnosed with a condition called Mayer-Rokitansky-Kuster-Hauser syndrome. In plain words, Christa had been born without a uterus. Arlette remembered saying to herself, "I wish I could let her use my uterus. I'll never use it again." Seven years later, Christa married Kevin Uchytil, and Arlette was ready. But the Mayo Clinic, when she asked for help, "turned us down," Schweitzer said. "They were worried about the controversy."[50] She soon found William Phipps, a reproductive endocrinologist at the University of Minnesota, who fertilized Christa's egg with Kevin's sperm and transferred the couple's embryos to Schweitzer's uterus. The twins—a girl and a boy and the genetic offspring of Christa and Kevin—were born that October.[51]

Schweitzer was willing to undertake the medical risks of pregnancy and childbirth at what was then considered the advanced age of forty-two because, after all, she was Christa's mother. This was a classic instance of what came to be called "altruistic" surrogacy. Arthur Caplan of the University of Minnesota called her decision "ethically admirable," but others were critical. Jay Katz of Yale University called Arlette's pregnancy "a very, very bad idea" that could create emotional difficulties for the children. Another ethicist, Albert R. Jonsen of the University of Washington, told a *New York Times* reporter that "surrogacy cuts to the heart of society's notions of who a person is and what constitutes a blood relationship."[52] Arlette and Christa shrugged off the criticism. When they were interviewed twenty-two years later, neither had regrets.[53]

Arlette Schweitzer was among the first cohort of gestational surrogates, and the nation was just beginning to get used to the idea that such women were not the genetic mothers of the children to whom they gave birth. The public still thought of a surrogate as a woman hired to give birth to her own genetic offspring and then hand the baby over to the couple who paid her.

This earlier form of surrogacy, which we now call "traditional," emerged in the United States even before the arrival of IVF, with surrogacy brokerages opening for business in 1978. The surrogate was hired—usually—by a married couple where the wife was unable to carry a pregnancy. The only technology involved was artificial insemination, which had been around for decades. A broker, typically an attorney, located the surrogate, who agreed, for a fee, to be artificially inseminated with the husband's sperm, carry the baby to term, and then relinquish the child. Estimates of how many babies were born this way varied wildly, and no reliable statistics were kept.[54]

The American public became aware of the practice only when the legal arrangements began to go awry, as they did in several places. The most sensational case occurred in New Jersey in 1985, and it came to define everything that was wrong with having babies for hire. Many Americans remember the famous "Baby M" case. Mary Beth Whitehead had signed a contract with William and Elizabeth Stern, agreeing to be inseminated with William's sperm. If she conceived and bore a child, their contract stipulated, she would be paid $10,000, and William and his wife Elizabeth would receive the baby. Whitehead, however, found herself unable to give up the baby after she was born, and the Sterns took her to court. Superior Court Judge Harvey Sorkow ruled that the contract was valid.[55]

Whitehead appealed. A year later, the New Jersey Supreme Court overturned Sorkow's opinion. Although awarding William Stern primary custody, it restored Whitehead's parental status, granting her all the rights of a noncustodial parent. The court condemned the practice of surrogacy in the strongest terms: "This is the sale of a child, or, at the very least, the sale of a mother's right to her child, the only mitigating factor being that one of the purchasers is the father. Almost every evil that prompted the prohibition of the payment of money in connection with adoptions exists here."[56] Baby M came from Whitehead's egg, Whitehead carried the pregnancy, and Whitehead gave birth. It seemed obvious to the court that no one else could lay claim to being the baby's mother, the contract she signed notwithstanding.[57]

The message of the Baby M case was that traditional surrogacy, where the surrogate is also the genetic mother, was morally problematic and created legal landmines. By then, however, the new reproductive technologies were able to separate genetics from gestation. In 1984, Wulf Utian, an infertility specialist at Mt. Sinai hospital in Cleveland, opened what he called a "host uterus" program, and his clinic became the first one to offer IVF to gestational surrogates in the United States. Two years later, its first baby was born. The genetic parents were Eliot, a cardiologist, and his wife, Sandy. In their thirties when they contacted Utian's clinic, they had been married since 1973. Sandy's diseased fallopian tubes had been removed, and the couple had believed they would never have children. But the advent of IVF gave them renewed hope. In 1982, they traveled to England to Bourn Hall, the IVF clinic established by Steptoe and Edwards. Their treatment was successful, and the couple eagerly awaited the birth. But the baby, born at twenty-eight weeks and weighing just 2 pounds, died, and complications during the delivery resulted in Sandy having to undergo a hysterectomy. She and Eliot did not want to use traditional surrogacy. "I didn't want the mother of my child to be someone else," Eliot told *People* magazine, "I wanted it to be my wife."[58]

Before finding Utian, the couple had been turned down by other IVF clinics. Worried that Utian would also reject them if they admitted to using a paid surrogate, they lied, saying that she was a family friend who offered to undergo the pregnancy for altruistic reasons. In reality, they had used surrogacy broker Noel Keane, who arranged for them to hire Shannon Boff. She agreed to undertake the pregnancy for a base fee of $10,000 plus travel and other expenses. In the end, their daughter Shira cost the couple about $32,000, including the $5,000 fee to Keane for his services. Boff, who had already been a traditional surrogate, said later that that this pregnancy posed fewer emotional problems. "With the first surrogate baby," who was her own biological offspring, "I had to keep telling myself that this was the couple's child," she recalled. "This one I knew wasn't mine. That made it easier."[59]

By the turn of the twenty-first century, Utian's program had treated one hundred and twelve couples, resulting in twenty-five live births.[60] He accepted only married heterosexual couples in which the wife was unable to carry a pregnancy, most commonly because of a prior hysterectomy. Just fourteen of the couples were using family members as surrogates, and most of the other ninety-eight used surrogacy agencies. Showing how fraught the idea of hiring a stranger to bear your genetic child was (and in many quarters remains), the clinic protocol called for all couples to undergo a psychological evaluation. That assessment was given to the IVF team, which had the final say on the "suitability" of the couple. Twenty-five live births in fifteen years, in the larger scheme of things, is not very many. There were actually a few more babies, though, because six of the pregnancies produced twins, and one, triplets.[61]

Within a few years, it became clear that gestational surrogacy was less objectionable to the public than traditional surrogacy. According to sociologist Susan Markens, traditional surrogacy evoked the idea of "baby selling," while gestational surrogacy was increasingly framed as a solution to the "plight of infertile couples."[62] The most powerful case for the latter argument occurred in California, where an African American woman, Anna Johnson, had agreed to bear the genetic child of Crispina and Mark Calvert. Crispina, a nurse, had lost her uterus to cancer surgery, but she did have functioning ovaries. The couple and their surrogate were treated by Ricardo Asch. (A few years later, they were among those suing the clinic for misplacing their embryos.)[63] The two women were acquaintances who worked for the same hospital. They agreed on a fee of $10,000 and had a California surrogacy center draw up a contract. The case soon took on racial overtones. Even though Crispina was Filipina, because she was married to a Caucasian the arrangement was often described in the press as a situation in which an African American carried a child for a white couple. After the baby was born in September of 1990, Johnson, like Mary Beth Whitehead, said that she had bonded with the baby during pregnancy and could not give him up. But the California Supreme Court decided that genetics was the determinative factor when it came to the definition of a "mother." Johnson was not the genetic mother of the baby; Crispina Calvert was.[64] The Calverts won.

Anna Johnson's feelings to the contrary, gestational surrogacy did not seem to involve the same moral dilemmas as traditional surrogacy did. At least some women who had been both traditional and gestational surrogates said that their experiences were significantly different when the egg that produced the baby was not their own.[65] This was an attitude especially prevalent in the United States. In England, for example, the woman who carried the pregnancy was considered the legal mother. This was the law when Bourn Hall became the first IVF program in the United Kingdom to offer gestational surrogacy in 1990, and it is the law today.[66] But in the United States, the general belief was—and is today—that the genetic parents are the "real" parents.

But if genetics takes precedence over gestation in the case of gestational surrogacy, the opposite view held true for egg donation. In gestational surrogacy, if the intended mother provided the egg, she was considered the mother, not the woman carrying the pregnancy. In Christa's situation, most people said, in effect "her egg, her baby." So too with Crispina Calvert. And that argument was also used by unhappy "traditional" surrogates when they either refused (as Mary Beth Whitehead did) to give up, or regretted giving up, their parental rights. In the United States, except for a few cases that have gone to court, if a gestational surrogate has a baby for a couple using that couple's gametes, both the surrogate and the couple consider the baby to be the couple's child, not hers.[67]

Gestational surrogates are not considered mothers because they do not provide the egg to create the pregnancy. But where does that leave egg donors? Here the situation is reversed, and the woman who carries the pregnancy, not the one who provides the egg, is considered the mother. Apparently, egg donors firmly and consistently believe that the woman who gives birth is the real mother. As one donor explained to sociologist Rene Almeling, voicing sentiments that were repeated over and over again by others, "Being an egg donor, it's not a tangible thing. . . . I mean, it [i.e., the egg] came out of me, but it's just like giving blood. You're giving something away and you don't see it again. It goes into someone else's body. It's gone." Equating egg donation and blood donation was a common refrain repeated time after time by the women whom Almeling interviewed.[68] Genetically speaking, a gestational carrier who is using the egg of an intended mother and a woman who has used a donor egg to conceive are in the same situation. Each is pregnant using someone else's egg. In genetic terms, neither is the mother of the baby she is carrying. In cultural terms, however, the two types of pregnancies are viewed as opposites. What matters is who intends to be the rearing mother. The genetic mother using a surrogate is the mother. The gestational mother pregnant with a donor egg, however, is also the mother. It may not be accurate genetically, but it makes perfect cultural sense.[69]

Egg donation provided new options for infertile couples. In a different way, so too did embryo freezing. With the development of effective cryopreservation techniques in the mid-1980s, couples could, if they wanted, freeze their extra embryos for later use or donate them to another infertile couple. The first birth from a frozen embryo occurred in Australia in 1984, and in 1986 Richard Marrs was the first to succeed in the United States. Once the effectiveness of the technique became evident, freezing extra embryos became a regular practice, but it remained a marvel well into the 1990s. One of us—Wanda—has never forgotten the first time, as a young obstetrician, she delivered a baby whose existence began as a frozen embryo. It was in the early 1990s, and after the infant was born, she and the patient's husband examined the baby and were amazed that she appeared to be perfect after being "frozen" for two years. She and the husband kept looking at the baby in wonder, ignoring the mother, who finally said, after a few minutes, "I'd like to hold my daughter now."

Freezing extra embryos gave couples a range of options. They could keep the embryos frozen until they wished to have another baby, at which time the they would be thawed and implanted; they could have them discarded; or they could donate them anonymously to another woman or couple. One study of such couples' decisions showed that 72 percent intended to use the embryos themselves at a later time, and nearly all had done that by the time of the study. Most of the rest chose to have the extra embryos destroyed. Just ten of the ninety-eight couples in the study said they were willing to donate their embryos, and in the end, only four couples actually did so.[70] The other six couples changed their minds. The authors of the study don't say why, but it is possible that as they watched their own babies develop, they realized that they didn't want anyone else bearing and raising their children's full genetic siblings. And if their embryos failed to provide them with children, they might not have been able to accept the idea of someone else having a child from their genes when they could not.

A separate study showed that couples were more likely to donate extra embryos created with donor eggs. In a group of sixty-eight women who underwent IVF with donated eggs (90 percent of them using the sperm of their husband or partner) between 1991 and 1996, 16 percent of them donated their extra embryos. The authors speculated that "relatively greater genetic distance" may have played a part in the decision—in other words, the fact that the wife was not the genetic mother made a difference. In any event, donated embryos accounted for a small proportion of births after IVF during the 1990s—we estimate approximately 2 percent in 1999.[71]

The success of embryo freezing also made possible a more worrisome development, the creation of premade embryos. Kathy Butler and her husband, Gary, had exhausted their savings, having spent $16,500 on a failed effort to conceive using an egg donor. So, when their doctor, Mark Sauer, offered them "ready-made embryos" for $2,750, they accepted immediately. As *New York Times* reporter Gina Kolata told her readers, the doctors at Columbia Presbyterian Medical Center in Manhattan "mixed human eggs and sperm to make a variety of embryos with different pedigrees, then froze them."[72] Sauer had these embryos created after one of his patients, who had contracted with an egg donor, changed her mind at the last minute. Such changes of heart did happen occasionally, he told Kolata, and with the donor's ovaries already stimulated, he went on, "It would be a waste of eggs" not to use them. Sauer fertilized somewhere between twenty to thirty eggs with sperm purchased from a commercial sperm bank, selecting for a variety of physical characteristics. Some of the sperm donors had blue eyes, others brown; some had blond hair and others brown or black. Patients could select for the greatest likelihood of physical resemblance to the woman's male partner if they wished.

The Butlers had five embryos implanted in Kathy's uterus. Two of them survived, and one of them split. In 1997, Kathy was pregnant—with triplets—at the age of forty-seven. "Premade human embryos are rare," Kolata wrote, and they were "largely confined to a handful of burgeoning centers like the one at Columbia Presbyterian, where doctors quietly tell patients about the embryos but do not advertise them." Kolata was concerned about what she called "a tangle of ethical issues, like the potential, in theory, for siblings to be raised by separate parents without any knowledge that they have brothers or sisters." But when she questioned Sauer about the ethics of the practice, he claimed not to see a problem and seemed to bristle at her implied criticism. "Don't accuse us of playing God," he told her.[73] But if Sauer or others continued the practice, it occurred in private. This was the only instance we found that reached the public eye in the 1990s.

Overcoming Male Infertility and Revitalizing Aging Eggs

In 1992, the development of intracytoplasmic sperm injection, called ICSI, changed the treatment of male infertility. Before ICSI, a woman whose male partner was infertile owing to deficiencies in semen quantity and quality had only a small chance of bearing that man's biological child. If she wanted to become pregnant, in some cases her only option was donor insemination, a practice dating back to the late nineteenth century.[74] By the 1930s, when advances in understanding the menstrual cycle made artificial insemination more effective, performing the procedure using the husband's sperm was becoming acceptable, but the use of donor sperm for male infertility remained in the shadows. Wives considering the use of donor sperm sometimes wanted the fact kept from their husbands, and many men, doctors insisted, would balk at the idea of creating a child who would be a daily reminder of that husband's inability to be a biological father. There were moral and religious objections to the practice, with the Catholic Church adamantly opposed.[75] But doctors had little else to offer. All they could do was find the couple a sperm donor and promise to keep their secret, a pattern that prevailed well into the 1980s, when the modern commercial sperm bank was born.[76]

Male factor infertility is the principal cause of a heterosexual couple's infertility in about a third of cases and plays a role in another third, but when IVF was developed, no one expected it to be useful in treating this problem.[77] Then, in the mid-1980s, despite the absence of reliable research, some clinics were employing it for male infertility, and by the end of the decade, two new techniques that involved the micromanipulation of eggs and sperm gained traction among some practitioners. One was called partial zona dissection (PZD) and the other subzonal insemination (SUZI). They enjoyed a brief flurry of popularity in some IVF clinics but quickly fell into disuse when studies found that they did not appreciably improve pregnancy rates in cases of male infertility.[78]

ICSI, however, was different. It really did work in many cases. In 1992, Andre Van Steirteghem and colleagues at Vrije Universiteit Brussel reported four births after ICSI in the medical journal *Lancet*. The technique involved the injection of a single sperm through the zona pellucida and into the egg itself. Subsequent studies demonstrated that ICSI was the first truly successful technique to make it possible for infertile men to become biological fathers. If a physician could find some normal sperm almost anywhere within a man's reproductive tract—it did not have to be in his semen—that sperm could be removed and used to fertilize his female partner's egg.[79]

ICSI could only be used with IVF, which meant that an otherwise fertile woman had to undergo IVF to become pregnant in this way, but to many heterosexual couples, the expense and the discomfort seemed reasonable prices to pay for the opportunity to have a child who was their genetic offspring. In the United States, the significance of ICSI became evident in 1995, when about 12 percent of IVF cycles were using ICSI. The authors of one report concluded that "the effects of male factor infertility are less evident than in previous years, suggesting that clinics with proficiency in ICSI may be able to mitigate the effects of male factor infertility." By the end of the decade, ICSI was used in about 43 percent of all IVF cycles.[80] Now, couples who could afford the procedure—or decided to go into debt for a child that was their genetic offspring—could conceive without using donor sperm.

ICSI transformed the treatment of male infertility. It did not put an end to donor insemination—it just made a shift in the major market for it. The newfound dominance of sperm banks after the AIDS crisis had put an end to the use of fresh sperm. It also created new opportunities for lesbian couples and unmarried women, who became the primary markets for the product.[81] Data on artificial insemination, because it is considered a medical treatment for infertility and not an assisted reproductive technology, are notoriously lacking. Indeed, no one knows how prevalent donor insemination is, nor is there reliable information on the number of babies born as a result. All we can say with any confidence is that sperm banks, to judge from their advertising and the numbers and range of donors they use, are doing a land office business.[82]

If medical science found a way for infertile men to have biological children, could it also make it possible for women whose eggs failed to fertilize to avoid having to use donor eggs? Despite increasing acceptance of egg donation in the 1990s, not every woman unable to conceive with her own eggs was willing to take this path to motherhood. Embryologist Jacques Cohen, one of the most prominent scientists in the field since the early days of IVF, believed he had figured out a way to make such women's eggs fertilizable. Cohen, at the time the director of the embryology laboratory at St. Barnabas Medical Center in New Jersey, hypothesized that he could use the cytoplasm (the part of the egg containing mitochondrial DNA) from a younger donor egg to "revitalize"—his word—aging eggs. He called this technique "ooplasmic transplantation." The press called the results "three-parent embryos."[83] Maureen Ott, who turned thirty-nine in 1996, did not care what they were called. She was interested. An engineer living in Pennsylvania, she had been undergoing cycle after cycle of IVF for seven years. She and her husband had almost given up hope when Cohen asked if she wanted to try a new technique. He said he could add what he referred to as "a small amount" of cytoplasm from the egg of a younger woman to Ott's egg before fertilizing it with her husband's sperm.[84] The Otts agreed to the procedure. It worked. Maureen conceived, and the couple had a healthy baby girl.

Cohen and his colleagues went about this experiment carefully. They received permission for the study from the institutional review board of St. Barnabas and made sure the patients were fully informed about what the procedure entailed and its experimental nature. One hundred couples were offered the technique. Thirty-three, including the Otts, accepted. Cohen, as he fertilized each of their eggs, injected the cytoplasm of an egg donor into the egg of the genetic mother. Because he believed that what he called "cytoplasmic deficiency" was an important reason for repeated implantation failure, he expected that replacing part of the cytoplasm of an egg from the intended mother with that from a donor would make it easier for these women to become pregnant. Thirteen of the women gave birth.[85] Most of the couples who had children remained anonymous. The Otts told their story publicly.

What was the difference between using a donor egg and using the cytoplasm from a donor egg? In the former, an entire egg from a donor is fertilized and transferred into the uterus, making the donor the genetic mother of the child. In the latter, only the cytoplasm from the donor egg was used. The nuclear DNA, the source of most of our heritable features, was Ott's. She would be the genetic mother. (Nuclear and mitochondrial DNA are described more fully in chapter 7, when we discuss contemporary techniques of mitochondrial replacement.) Most of the couples who participated in Cohen's study had previously declined to use donor eggs. But a donor's cytoplasm was different and seemed trivial to them. Soon after their births, eight of the thirteen babies were tested for the donor's mitochondrial DNA. Only two of the babies had retained any of it.[86]

St. Barnabas was not the only IVF center experimenting with the technique. The Jones Institute was also trying it, announcing its first pregnancy in 1998.[87] Other programs may have done so without drawing attention to themselves. There were an estimated thirty children born using this technique; only two—Emma Ott and Alana Saarinen—were ever publicly identified. Sharon Saarinen, Alana's mother, was not part of Cohen's study. She was treated in Michigan, where she said her doctor "explained to me in layman's terms that it would give my egg a boost and . . . make it more likely that we could become pregnant." She said she didn't remember being told that the procedure was experimental, but if so, it wouldn't have mattered. "Even if there had been risks . . . it wouldn't have stopped me. I just wanted a child so very much."[88] However many IVF programs were using it, they were all forced to stop after Cohen and his research group published an article in *Human Reproduction* in 2001 reporting on their success at St. Barnabas. The only researchers to publish their data, they caught the attention of the US Food and Drug Administration, which concluded that the procedure was a form of genetic modification.[89] The agency determined that the use of a third person's cytoplasm in IVF therefore required the filing of a new drug application

(NDA) for its review.[90] St. Barnabas originally agreed to proceed with the NDA, but it abandoned its efforts two years later, Cohen said, for "lack of funding."[91]

From Wyden to Dickey-Wicker to Efforts at Self-Regulation

Soon after the Wyden Act was signed into law by President George H. W. Bush, Democrat Bill Clinton was inaugurated. His victory had raised hopes in the biomedical community that he would be more open to funding research on IVF and human embryos than his two predecessors had been. Early indications favored those expectations. Clinton appointed the distinguished cancer researcher Harold Varmus as head of the National Institutes of Health. Varmus in turn quickly impaneled the Human Embryo Research Panel, which included scientists, physicians, ethicists, and patient advocates. Completing its work in September of 1994, the panel recommended an end to the federal funding ban. Its detailed report provided guidelines for federal funding of research on existing embryos and for the creation of embryos specifically for research.[92] In December, the Advisory Committee to the Director of the NIH unanimously endorsed the report. After twenty years, it appeared that the federal funding drought had ended.[93]

It took only hours for those hopes to deflate. Clinton almost immediately overruled the panel's recommendation to allow the creation of embryos specifically for research. As Ronald Green, a member of the panel, wrote dispiritedly, "the President's edict seemed to be a repudiation not just of the Report, but of the long process that led up to it. It appeared that once again in the area of reproductive medicine, political considerations—in this case, pressures created by the recent election of a conservative Republican Congress—had preempted the work of a specially appointed panel of advisers." But as disappointed as he and the rest of the panel were that Clinton rejected this specific recommendation, Green added, the new rules would enable the NIH to fund research on "improving techniques of in vitro fertilization, basic research using 'spare' embryos remaining from infertility procedures" and related investigations.[94] It turned out that Green was too optimistic. The Republican Congress responded to this new report by attaching a rider to the next appropriations bill prohibiting federal funding of any research that created a human embryo for research purposes or destroyed an existing embryo.

That rider, referred to as Dickey-Wicker for the two legislators who introduced it, enshrined the 1975 federal funding moratorium into law. Technically, it only applied to the federal budget for the 1996 fiscal year. But the same amendment has been passed in every subsequent budget. Clinton signed the measure in 1995, as has every president since, effectively negating the work of the committee and keeping human embryo research in the private sector.[95]

Another attempt to create consistent policies to guide the ongoing development and use of the new technologies came from the joint efforts of the American Fertility Society (which became the American Society for Reproductive Medicine, or ASRM, in 1994) and the American College of Obstetricians and Gynecologists (ACOG), which created the National Advisory Board on Ethics in Reproduction (NABER) in 1991. Their idea was to have this board act as "the private sector's answer to a presidential commission" on assisted reproduction, or perhaps "the nation's IRB [institutional review board] for reproductive medicine." ACOG and AFS provided the initial funding, but the organization floundered. Never adequately resourced, it seemed both too independent of its funding sources—sometimes promulgating opinions that differed from the official views of both professional organizations—and not independent enough, since it relied on them for survival. It tried and failed to secure other funding, and ultimately it was unsustainable. It disbanded in 1998.[96]

Once NABER fell apart, there was nothing else to take its place. Toward the end of his first term, President Clinton created a National Bioethics Advisory Commission. On the face of it, such a commission might have been perceived to be doing the same work as NABER, except with more governmental legitimacy. The commission's original charge was to "identify broad principles to govern the ethical conduct of research, citing specific projects only as illustrations for such principles." But in practice, the ethics board avoided questions about human embryo research, limiting its scope to the "protection of the rights and welfare of human research subjects; and . . . issues in the management and use of genetic information including but not limited to human gene patenting."[97]

The decade ended as it began, with no clear federal guidance on technologies that would only grow more complex and be more widely used. By the year 2000, the assisted reproductive technologies were fully embedded in the practice of reproductive medicine. Louise Brown, the world's first IVF baby, turned twenty-two that July, and Elizabeth Jordan Carr turned nineteen in

December. That year alone, some thirty-five thousand babies were born in the United States through these technologies, nearly 1 percent of the four million babies born in the country that year. Worldwide, between the time of Louise Brown's birth and the end of the first year of the new millennium, some sources estimated that about nine hundred thousand babies had come into the world as a result of assisted reproduction.[98] The result of all these changes was a shift in the contours of reproductive medicine. The field became more commercialized, new technologies were developed, and new uses were found for existing ones. More changes were to come, as the demand for assisted reproductive technologies grew, and the field of reproductive medicine moved into treatment realms that went beyond infertility.

7

BEYOND
INFERTILITY

"Are You as Fertile as You Look?" That provocative 2011 headline in the *New York Times* was aimed squarely at thirty-something, childless women. "Forty may be the new 30," reporter Tatiana Boncompagni wrote, "but try telling that to your ovaries." Women who had done everything they could to look younger than their years, she went on, apparently believed that good health and good looks, the typical markers of "youth and beauty," were indications not just that they had the money to purchase beauty products, take advantage of cosmetic dermatology, eat healthy food, and make time to exercise, but also of their ability to reproduce. Many of these women were stunned to discover otherwise. As a forty-one-year-old magazine editor put it, "I'd based a lot of my self-worth on looking young and fertile, and to have that not be the case was really depressing and shocking." She recently finished her fifteenth unsuccessful round of IVF, she told the reporter, and she and her husband were now looking for an egg donor and a surrogate. Another woman, a forty-five-year-old artist, said in frustration, "I watch what I eat, I don't drink, I take extremely good care of myself and I come from a very fertile family. . . . Everyone in my life told me I look young for my age." She believed, she went on, that "it was the same on the inside as it was on the outside."[1]

They felt blindsided, just as Melanie had. A high achiever from a working-class background, Melanie realized early on that intelligence was a necessary but not sufficient ingredient for success. Tenacious, disciplined, and holding an MBA from a first-rank university, by her late thirties she had become an executive at a nonprofit organization, rising to second-in-command and ambitious to hold the top job. She and her husband had always expected to have children someday, but for a long time, focused on their careers, they had never found what she called a "natural opportunity" to have a baby. Then she turned thirty-nine, and a sense of urgency set in. After the couple failed to conceive on their own, Melanie made an appointment with her gynecologist, who prescribed fertility drugs. When they did not help her, the next step was intrauterine insemination with her husband's sperm. That also failed, and Melanie and her husband sought more specialized assistance at a fertility clinic. They tried IVF twice, and it hadn't worked. Now, the couple was trying to figure out what to do next. Unaccustomed to failure, Melanie viewed her inability to become pregnant as just that—her failure. "There hasn't been anything I've wanted that I haven't gotten," she told journalist Liza Mundy. Until now.[2]

Melanie was not alone. She was part of a cohort of educated women who had decided to wait for career success or stability before thinking about getting pregnant. Whether it was being offered the top job, making partner, finishing residency and beginning a medical career, or earning tenure, many women whose professions required long years of preparation for success postponed parenthood until they felt secure enough in their life's work. But not every woman was exactly like Melanie. Other women simply did not find a life partner until their thirties. Carolyn was thirty-four when she and Barry married. They had met in Vermont, and after the wedding they settled near her family in New York City, where she worked as a fundraiser and he managed a restaurant. The couple had wanted children right away, but four years and multiple treatments later, they had not succeeded. The stress was undermining their marriage.[3] Carolyn, like Melanie, sought help from her gynecologist, and she too was prescribed fertility drugs and after that, intrauterine insemination with her husband's sperm. A fertility specialist, who advised IVF, came next. They couldn't afford it and were grateful when Carolyn's parents offered to cover the cost. But IVF failed as well.

Like Carolyn, Didi was in her mid-thirties when she married Mark. The couple lived in a small Pennsylvania city, where she worked as a sales representative and he was a landscape designer. After they married in 2008, she immediately began trying to conceive, expecting no difficulties. Her mother and sister had both gotten pregnant easily, and she was sure she would too. After a year without success, she saw a gynecologist, ultimately learning, she said, that she suffered from premature ovarian failure. Told that she was extremely unlikely to conceive—at least not with her own eggs—she was devastated.[4]

At some point during treatment, each of these women's doctors suggested they might want to adopt or use a donor egg. Deeply unsettled by the idea of carrying a pregnancy initiated with another woman's egg, Didi was taken aback by Mark's anger over her refusal to "bear his biological child." Your child, yes, she told him, but not mine. How could he expect her "to carry a baby that doesn't have my genes?" Still, she said, genes weren't her only concern. Didi's family had come from India, and she wanted a child who shared her ethnic background. Given both her reluctance to use a donor and the difficulty they were told to expect in locating one of Indian heritage, Mark ultimately came around to her way of thinking, and the couple adopted an eighteen-month-old boy from India. The adoption cost the couple $25,000, for which they took out a second mortgage on their house. For Didi, the cost of assisted reproduction had not been the major issue. They would have gone into debt either way. It was the idea of using a donor that troubled her.[5]

Melanie was equally uncomfortable with the idea of egg donation and told her husband that she wanted to adopt. His response was much like Mark's. "I can't believe you wouldn't want to at least try to have a child that was ours," he told her. Surprised and hurt, Melanie shot back that the baby would be his, not hers. If she became pregnant, she said, "I would feel like a fraud. . . . This wouldn't be my biological child, so why am I going through this whole charade?" For his sake, however, she gave in, only to have their selection of an egg donor fall through. Now they were both in doubt, trying to decide whether to look for another donor. Carolyn, in contrast, was completely unfazed by the idea of using a donor. She wanted a child and didn't care how that child came into her life, whether through adoption or egg donation. Her husband, however, was so unhappy about their difficulties that he was incapable of helping to make a decision. In the end, they used a donor egg. Carolyn decided that it was important to her to experience pregnancy and childbirth. "The baby will have the genetic link to Barry," Carolyn said, "but I will be the mother." She conceived during her second IVF cycle and gave birth to a baby girl.[6]

Similar stories were playing out in households across the country in the first decade of the twenty-first century. Ambitious professionals took center stage in the pages of books, columns, and articles in newspapers such as the *New York Times*, where the focus was on difficulties faced by women like Melanie, high achievers who wanted to have both a career and a child. Carolyn and Didi, however, whose stories were told in the *Ladies' Home Journal* (*LHJ*), were far more representative of the women and couples filling the waiting rooms of fertility specialists around the country. Few magazines were more middle-American than the *LHJ*, a popular magazine reaching millions of households every month.[7] Founded in 1883, it had a reputation for addressing uncomfortable family dilemmas affecting its readers, with solutions to those dilemmas that leaned heavily on the importance of maintaining conventional familial roles.[8] Now here it was, in its popular "Can This Marriage Be Saved?" feature, telling its readers that turning to egg donation was simply another way to create a family. Donor eggs, it appears, were becoming mainstream.

This is the first of two chapters bringing the story of assisted reproduction into the present. In this chapter, we ask how the field of reproductive medicine has changed—and not changed—in the twenty-first century. Melanie, Carolyn, and Didi were in some ways much like the women who wrote heartfelt letters to John Rock, waited vainly for Pierre Soupart's early IVF study to be approved, or applied to the Jones Clinic in hopes of being the first in the country to have a test-tube baby—a term still used by the public as a synonym for in vitro fertilization. But in other ways, they were part of a new wave of women and couples who, sometimes to their surprise, found themselves wondering if they should have made parenthood a priority before now, if a pregnancy with someone else's gametes would satisfy their parental longings, if they should have that fifth round of IVF, or if they should consider adoption. And many of the medical practices to which they turned were different as well. It was possible, even in the twenty-first century, to find physicians like Leonore Huppert at Pennsylvania Hospital or Steven Sondheimer at the University of Pennsylvania, longtime IVF practitioners with expertise in both reproductive medicine and general gynecology, but they were becoming few and far between. By the early twenty-first century, an IVF patient most likely would be treated at a fertility center—many of which were large private enterprises—specializing in IVF and other assisted reproductive technologies.

From Academic Medicine to the Fertility Industry

Popular media accounts of the new reproductive technologies and those seeking them in the past decade or so generally account for the growth in the field by saying that infertility has been on the rise and might even be "skyrocketing."[9] That's not true. Social and cultural changes, aided by improvements in the technologies themselves, not a rise in infertility, have been the most important forces driving the expansion of infertility services. The most important factor, but not the only one, is the greater willingness of hetero-sexual couples—mostly in their thirties but some in their twenties—to seek medical help when they fail to conceive. In addition, new markets for repro-ductive services have burgeoned—including women in their forties and even fifties, same-sex couples, and couples who are not infertile but are having their embryos diagnosed for specific genetic disease or screened for chromo-some abnormalities or sex selection.

In 2015, there were 499 centers offering assisted reproductive technolo-gies in forty-nine of the fifty states—all but Wyoming.[10] Fertility treatment had become a $3–4 billion enterprise. California, with 77 fertility centers, had the highest number, but the largest ones were on the East Coast. Four fertility centers in the country performed more than five thousand cycles a year, all of them for-profit operations and three of them located along the Washington, DC–New York corridor. The other was in Chicago.[11] Shady Grove Fertility, founded in the early 1990s in a Maryland suburb of Wash-ington, is the largest practice by far, with twenty-nine offices in Maryland, Virginia, Pennsylvania, Georgia, and the District of Columbia. In 2015, its clinics initiated 11,265 IVF cycles, almost twice as many as the next largest center. The dramatic growth of Shady Grove highlights a major shift in the field, with for-profit clinics becoming the nation's largest and fastest-growing providers of fertility services. In terms of volume, they have outstripped the academic medical centers.

When IVF came to the United States, the doctors who performed the procedure were faculty members in departments of obstetrics and gynecol-ogy at academic medical centers. Whether established experts in reproduc-tive medicine or just out of fellowship, these men and women had been fueled by the excitement of discovery, the satisfaction of helping a group of patients who had all but given up hope for a baby, the challenge of mastering an exacting new technique, the thrill of crossing a new reproductive frontier, and the desire to create a lasting legacy. Alan DeCherney, reflecting on his career since those early years, recalled spending an afternoon with Luigi Mas-troianni in 2008, during the older man's final illness, just days before he died. While they were together, the two men spent some time talking about Mas-troianni's mentor John Rock, and DeCherney reflected on how the arc of his own career unfolded as part of a much larger history. "It was like Rock passed the torch to him," he recalled, "and he passed the torch to me. It was very moving."[12]

Only a few of the men and women working on IVF in the early 1980s could trace their professional lineages back to Rock, but all of them were a part of the larger web of academic medicine. Every one of the first half dozen or so physicians reporting a live birth from IVF were practicing in an academic medical center. Today, among those medical schools, only a few preserve the hallmarks of that earlier period. Of the institutions in which live births had been achieved by 1983, just two—the Jones Institute at Eastern Virginia Medical School and Penn Fertility Care at the University of Pennsylvania—retain powerful and overt connections to their roots. The Jones Institute proudly bears its founders' names. At the University of Pennsylvania's medical school, the legacy of Luigi Mastroianni looms large, and its current division chief of reproductive endocrinology and infertility, Christos Coutifaris, is a Mastroianni protégé. But these two institutions are anomalies. At the University of Texas at Houston, the leadership of the Department of Obstetrics and Gynecology made a deliberate decision to deemphasize IVF in the mid-1980s. When this happened, Martin Quigley, who directed the program, moved out of state for another position and soon left the field of reproductive medicine altogether. In 1998, Vanderbilt ended its program as well. At both medical schools, it wasn't lack of expertise on the faculty but medical school priorities that resulted in their IVF programs moving from medical school departments of obstetrics and gynecology to private practices.[13]

We are not saying that academic medicine is no longer an important force in assisted reproduction. It surely is. The Jones Institute continues to flourish, and Penn, Yale, and the University of Southern California all have highly regarded IVF programs, as do many other medical schools. Furthermore, academic research fueling the ongoing development of assisted reproductive technologies and advancements in practice is conducted in departments of obstetrics and gynecology that have divisions of reproductive medicine. Academic medical practices are, however, no longer the largest providers of clinical services.

Freestanding IVF clinics were made possible by advances in technology that eliminated the need for access to a hospital operating room, but the rapid growth of the field also played a role. As more young physicians wanted to specialize in this area in the late twentieth century, Richard Marrs told us, new fellowship programs opened, and the numbers of trained reproductive endocrinologists ultimately grew beyond the capacity of academic medicine to absorb them. As Marrs remembered, "After a while there were not enough academic programs for the people who wanted to do IVF. Most academic centers had their programs and REIs [reproductive endocrinologists] didn't want to just join and be low down on the totem pole in the IVF program." As a result, by the early 1990s, "more people started setting up private centers." Marrs tried to exist in both worlds as long as he could, leaving academic medicine slowly and reluctantly. But leave he did. "I finally moved out of academics in 1992," he said. Even then he tried to keep his hand in, conducting basic research until 1997 and clinical trials for a while after that.[14] It is also true that from the perspective of deans and chairs at medical schools, who make decisions about academic priorities, faculty members can offer appropriate teaching in reproductive endocrinology to residents even without their own clinical programs in assisted reproductive technology. Residents who want to move on to fellowship training in the field, of course, need an academic program, but there are enough thriving fertility centers at medical schools to meet these needs.

Twenty-first-century reproductive medicine has flourished in the private sector, even as substantial numbers of physicians in many other fields have become employees of health systems, because patients generally pay for IVF and other advanced technologies out of pocket, and IVF centers can be highly profitable. Assisted reproductive technologies are more expensive in the United States than in other developed countries. Pricing is driven by the market even at academic medical centers, and some of them strongly resemble their for-profit counterparts. The fifth-largest IVF center in the country—and the only one of the ten largest programs in the United States that is located at an academic medical center—is the Center for Reproductive Medicine at Weill Cornell Medical College in New York City.[15] A few of its faculty earn the kind of head-spinning incomes that are usually associated with private enterprise. Cornell reported the compensation for the center's director as $6.3 million in 2016, and two of its other physicians were in the multimillion-dollar range as well.[16] Reproductive medicine pays well, but it does not typically pay *that* well. The average income for a reproductive endocrinologist in the United States in 2017 was just under $334,000.[17]

Debora Spar, a leading business scholar, suggested in 2006 that society needed to accept the fact that in the United States, reproductive medicine had become a business. Rather than moralize over a trend that is unlikely to be reversed, she said, the nation should regulate it like one. "Harsh as it may seem," she argued, "we need to view reproductive medicine as an industry, with all the commercial prospects and potential foibles that other industries display."[18] Spar, it turned out, was prescient. Today, investment bankers produce reports on "the fertility services sector" and "fertility market overview[s]" for potential investors, and private equity companies are putting their money into the for-profit fertility market. Thus far, there are no national chains of fertility clinics on the model of urgent care centers, but they may be on the horizon. William Schoolcraft, Colorado's first IVF practitioner and founder of the Colorado Center for Reproductive Medicine, joined forces with a Boston investment firm in 2015 for that purpose. His chain of practices, now known as CCRM, had branches in nine cities across the United States and one in Canada in 2018, and it is seeking to create branches nationwide.[19]

Infertility and Assisted Reproductive Technologies: The Complexities of Supply and Demand

Go to the website of many fertility centers in the United States, and you will see a statistic claiming that the United States experiences an infertility rate of 12 percent or more. This is an overstatement. The infertility rate in 2010, the last year for which comprehensive statistics are available, was 6 percent. In 1982, when IVF was just being introduced into the United States, the infertility rate was 8.5 percent. Translating these percentages to numbers, in 1982 there were about 2.4 million women experiencing infertility, and in 2010, there were about 1.5 million. But infertility statistics may not tell the whole story. There are other women—some with infertility, some not—who may have difficulty at some point in their reproductive lives either in becoming pregnant or carrying a pregnancy to term. Their problems may be transient or of longer duration. The National Center for Health Statistics places these women into a single category, describing them as having experienced "impaired fecundity." About 12 percent of women are or have been a part of this broader and more amorphous cohort. That's the same percentage as in the 1980s, but it represents a decline from a high of 15 percent in 2002.[20] More awareness of the need for early treatment of sexually transmitted infections may have played an important role in the decline in both rates. In 2002, the criteria for diagnosis of pelvic inflammatory disease were broadened, leading to new guidelines that resulted in more women receiving treatment, thus decreasing the risk of tubal damage and subsequent infertility.[21]

Although it is impossible to determine the precise overlap between the above statistics and doctor visits to discuss fertility problems, we do know that approximately 12 percent of women of reproductive age in 2010 had seen a doctor for medical help to have a baby. Some of them were having difficulty becoming pregnant, others with carrying a pregnancy to term. The most common help they received was advice, for example, on the frequency and timing of sexual intercourse. Others received a range of medical treatments, and slightly less than 1 percent (0.7 percent) underwent treatment with assisted reproductive technologies. While that may seem like a small number, it represents a sevenfold increase from fifteen years earlier.[22] Overall, according to RESOLVE, about 65 percent of people treated for infertility succeed in having a child. Most of them, about 85 to 90 percent, undergo medical treatments such as ovulation induction, intrauterine insemination, and surgical procedures to have their babies. Only about 3 percent require assisted reproductive technology to conceive.[23]

The cultural significance of the assisted reproductive technologies, however, extends beyond the relatively small numbers of Americans who use them. In vitro fertilization and all its contemporary permutations and combinations receive a great deal of publicity, and that publicity calls attention to the fact that infertility treatments exist, and that they can be effective. Infertility success stories in magazines such as *Real Simple* or *Ebony* attract readers from diverse groups, and there are hundreds of blogs and online communities that appeal to multiple audiences. RESOLVE offers a wealth of information and access to support groups, and private fertility centers advertise extensively. Between 1999 and 2015, the volume of treatment in America's fertility centers, measured by the number of egg retrieval cycles, went up more than two and a half times, to nearly 232,000 retrieval cycles, about 80 percent of them with the intent to achieve a pregnancy, and the rest for the purpose of freezing and banking the resulting embryos for future use. Just under 61,000 women gave birth after being treated that year, for an overall take-home baby rate of about 33 percent, up from about 25 percent in 1999.[24] Most of the patients in 2015, like those in years past, were under the age of forty and using their own eggs and their partners' sperm in their attempts to conceive.[25] (See figures A.1 and A.2 for additional information.) In 1999, 90 percent of retrieval cycles had involved heterosexual couples using their own eggs and sperm. In 2015, it was 87 percent.[26]

Success rates for women aged forty-two or younger rose in the early twenty-first century; younger women, as would be expected, were the most successful. By 2012, women aged thirty-five or younger had a 50 percent live birth rate, up from about 35 percent a decade earlier. Women aged forty-three and older fared the worst, at least when they used their own eggs.[27]

Today, women in their thirties (and some in their twenties) continue to represent the largest national market for infertility services, but there has been a significant increase in the numbers of women in their forties seeking treatment. Surprised to discover that their age meant that they would have more difficulty conceiving, or praying to be the exception to the rule, such women hope that medical science can make it possible for them to become mothers. In 2000, 13 percent of cycles at the nation's fertility centers had involved women who were forty or older. In 2015, such women accounted for 22 percent of patients.[28] In cities like New York, Los Angeles, and San Francisco—metropolitan hubs that have significant populations of women with careers, or money, or both—the percentages are higher. About 45 percent of the patients at California Fertility Partners in Los Angeles and at New Hope Fertility Center in Manhattan in 2015 were forty and older, as were 30 percent or more at the fertility centers at Weill Cornell Medical College, Columbia University, New York University, and the University of California, San Francisco. In contrast, in most areas of the country, the patients are younger. In Minneapolis, at Minnesota's largest IVF center, 48.5 percent of the patients were under thirty-five, and just 14 percent were over forty. In Philadelphia, at Penn Fertility Care at the University of Pennsylvania, 45 percent of patients were under thirty-five, and 17 percent were over forty.[29]

The trend toward later childbearing, which is especially evident among women with college degrees, means that such women are not likely to discover that they or their partners have a fertility problem until they are older than their counterparts in earlier generations. At the turn of the twenty-first century, only 42 percent of college-educated women between their mid-twenties and mid-thirties had borne children. But if we look at what happened to this exact same cohort of women by 2010, now that they were thirty-five to forty-four years old, about 76 percent of them had children. These women were still having babies; they were just having them later in life. In fact, among all women, 85 percent had borne at least one child by the time they were forty-four.[30] Most of these women, but by no means all of them, can have those children without medical intervention. About 30 percent of all first births in this country were to women over the age of thirty in 2015—21 percent to women between thirty and thirty-four, and 9 percent to women over thirty-five.[31]

Waiting longer to marry and bear children is perhaps inevitable in a society in which women bear the brunt of marriage's domestic expectations and where motherhood (but *not* fatherhood) adversely affects almost every aspect of a woman's working life.[32] Truly egalitarian gender roles and expectations in American society have yet to materialize and do not appear to be on the horizon anytime soon, and one way to cope with this situation is to delay marriage and childbearing. In 1960, when many women held jobs but few had careers comparable to those of men, women married on average at the age of twenty-one and men at twenty-four. Fifty years later, the age of first marriage for women was up to almost twenty-seven, and for men it was nearly twenty-nine.[33] Although some observers look at these later ages at marriage and conclude that marriage as an institution is in trouble, that does not seem to be the case—at least not yet. People are still marrying in significant numbers. By the time they reach the age of forty-five, about 90 percent of men and 92 percent of women have been married at least once.[34]

In the 1950s, marriage had been a signal that a couple was embarking on their adult roles. Women had their first child in their late teens and early twenties, and by the time they were in their early thirties, many had completed their families. Today, as historian Stephanie Coontz puts it, marriage is likely to be "the final rather than the first step in the transition to adulthood," and later childbearing is a consequence of this trend.[35] Delayed childbearing has its hazards, however. Fertility in women, medical experts say, "decreases gradually but significantly beginning approximately at age 32 years and decreases more rapidly after age 37 years."[36] Today, couples are likely to take action when a pregnancy does not arrive when they hoped it would.

If the increased use of infertility services among couples in their thirties is the predominant reason for the growth of assisted reproduction in the twenty-first century, it is not the only one. New categories of patients are also expanding the market for fertility services. They include women in their mid-forties and older, nearly all of whom will require donor eggs to conceive; same-sex couples; single women; and some single men. In addition, fertility centers treat patients who are employing IVF for reasons other than infertility. Couples in which one or both is a carrier of a genetic disease can have their embryos undergo preimplantation genetic diagnosis, ensuring that only those embryos without the genetic defect are implanted. Other couples may want their embryos to undergo genetic screening to rule out chromosome abnormalities, and some desire a baby of a particular sex. Another group of patients includes young women, from their teens to their early thirties, about to undergo medical treatments that could compromise their future fertility. These women are freezing embryos, eggs, and, among the very young, ovarian tissue. Egg freezing has also attracted the attention more recently of women who are delaying pregnancy for personal reasons. We circle back to this phenomenon, called elective or "social" egg freezing, later in the chapter.

Another expanding patient category includes couples whose inability to procreate may not be due to medical factors but who nevertheless are unable to conceive without medical intervention. Lesbian couples cannot conceive without medical assistance because they do not have sex with men. They require a sperm donor. If there are no other underlying fertility issues, these couples, as well as unmarried women without a partner, can usually conceive with intrauterine insemination using donor sperm, a medical technique offered by general gynecologists as well as reproductive endocrinologists at specialized fertility centers. In the event that one partner wishes to provide the egg and the other partner to carry the pregnancy, the couple will require IVF. A same-sex male couple has an analogous situation, since neither partner has ovaries or a uterus. If the couple wants genetically related children, they will need both donor eggs and a gestational surrogate. Such services are easy to find in many parts of the country; nearly every fertility center in or near a major metropolitan area provides services to same-sex and transgender couples, and most of them offer services tailored to their specific needs. The growth in these new patient cohorts, from postmenopausal women needing donor eggs to same-sex couples and fertile couples using assisted reproduction for other reasons besides becoming pregnant, has also helped to spur an increased use of assisted reproductive technologies in the twenty-first century.

Last-Chance Babies

In 2015, the *AARP Bulletin* published a story called "Last Chance Babies." Yes, the AARP. The major lobbying association for older Americans in the United States seems to have concluded that postmenopausal childbearing was relevant to the lives of its members. The article profiled several women who had babies in their fifties and sixties, among them Sarajean Grayson, who delivered a son at the age of fifty-one and twins at fifty-three. She had three children from a prior marriage, but her current husband, thirteen years her junior, had never been a father. "Becoming a mother again was all about giving her husband, a former Roman Catholic priest, a profound gift of love," wrote the reporter. Sarajean added, "My husband is a good man [and] I wanted him to have the experience of parenthood more than anything." And Sarajean wasn't the oldest of the mothers profiled. New Jersey psychologist Frieda Birnbaum was already a mother of two children when she had a third child at fifty-three. Then, at the age of sixty, she became pregnant again, giving birth to twins. Frieda had been turned down by several clinics in the United States, she said, who refused to accept a sixty-year-old. She and her husband decided to travel to South Africa, where "the youngish looking blonde lied about her age." When reporters came to talk to her about the birth, she said they looked surprised. "They thought I was supposed to be some shriveled up, gray-haired woman."[37]

Sarajean and Frieda could not have conceived with their own eggs. A woman who has a baby in her fifties today, unless she has high-quality unused frozen embryos or, in a scenario that is much less likely, fertilizable frozen eggs, requires a donor. The two of us have studied fertility and infertility for more than a quarter of a century, and we found ourselves surprised, in researching this book, at the level of public misunderstanding of age and fertility. We were sure that most people understood that the middle to late forties marks the end of virtually every woman's reproductive years. We were wrong. A sizeable portion of the American public apparently believes that conception is possible at almost any age. Nearly three-quarters of respondents to a recent poll agreed with the statement "that looking and feeling good and being healthy overall was as important as biological age when trying to conceive."[38] This is simply not true. In America's IVF clinics in 2015, just 1 percent of patients aged forty-five and older were able to have a baby using their own eggs, and the odds weren't significantly better for younger forty-somethings. They were 3 percent for women aged forty-three and forty-four, and just 8 percent for women aged forty-one and forty-two.[39]

Americans are riveted by stories of celebrities and other well-known figures becoming pregnant or having babies in their fifties. Most of these women deflect questions about the origin of the eggs that made those pregnancies possible, helping to fuel the notion that their babies came from their own eggs. Of course, any woman, public figure or next-door neighbor, has every right not to reveal how her pregnancy occurred. Still, whenever any of these women mention having had some unspecified form of IVF without noting that the procedure included donor eggs, or perhaps embryos created with her own eggs and frozen years earlier, that pregnancy produces a perception in the society that women can have a baby at fifty-two or fifty-nine without resorting to a donor or a previously frozen embryo.

Whether arising from all the publicity given these late-in-life babies or just a general refusal of many people to admit that forty or fifty is not the new thirty, it apparently comes as a surprise to many of these women that they are indeed too old to get pregnant with their own genetic material, and attempts by the reproductive medicine establishment to enlighten them have generally fallen flat, or worse. In 2001, the American Society of Reproductive Medicine funded a public service advertising campaign using posters featuring the various factors that can affect fertility. Their first three in the series warned that smoking, excessive weight, and sexually transmitted infections could all harm fertility. Then came the fourth poster, featuring an image of a baby in an hourglass, captioned "Advancing Age Decreases Your Ability to Have Children." To reproductive specialists, this seemed to be just as much of a straightforward statement as the other three, a message that might prompt a visit to the gynecologist. Instead, it ignited a firestorm. Women saw it as an attempt to frighten them into having children at a young age. The campaign backfired.[40] Gynecologists and reproductive endocrinologists have not given up on making the point that aging women have aging eggs and that aging eggs have increasingly lower chances of fertilization. There has not, however, been another ad campaign. Instead, there has been a growing market for donor eggs.

Once it became clear that, in terms of achieving a pregnancy, what matters most is the age of the donor, not the age of the woman seeking to become pregnant, fertility clinics began seeing more women in their early to midforties.[41] If a woman between the age of forty-two and forty-seven uses an egg donor and is able to have one embryo transferred into her uterus, her likelihood of taking home a baby ranges from close to 60 percent at the younger end of the spectrum to 50 percent at the higher end.[42] Achieving a pregnancy in the late forties, or even fifties, has become easier with donor eggs. Older women, however, especially those over fifty, face an elevated risk of serious complications during their pregnancies. For the most part, however, these problems are not dealt with by the fertility clinic making it possible for these women to conceive. They are left to the obstetricians and maternal-fetal medicine specialists who care for women with high-risk pregnancies. The reproductive endocrinologists who make those pregnancies possible are not the ones who provide prenatal care, manage the complications, and deliver the babies.

The process of acquiring donor eggs has been simplified in recent years. New and more effective methods of egg freezing, which we describe later in the chapter, are making possible the creation of egg banks built on the model of sperm banks; some sperm banks have added egg banking to their businesses. Cryos USA, the American branch of one of the oldest and largest international sperm banks, now offers an online egg donor catalogue that is almost the mirror image of its catalogue of sperm donors. It includes the height, weight, interests, and level of education of the donors. There are photographs as well; some donors protect their privacy by including only baby pictures, but others include current ones. Customers choose their egg donor and see the price—for "Ariana" in 2017, the cost was $5,000 for two eggs, with a recommended minimum purchase of six to eight eggs to have the best chance of achieving two embryos. Eggs are added to a "basket" for purchase. Once the transaction is completed, Cryos ships the eggs directly to the IVF center where they will be used.[43]

This direct-marketing approach, while it may be easier, is not universally embraced by existing egg donor agencies, many of which have had a longstanding practice of matching individual donors and recipients and providing counseling for both as a part of the process. But what will happen to that older model of egg donation now that prospective parents can simply choose from a catalogue? Egg donation has traditionally been cast as a gift exchange: for the privilege of using another woman's egg to bear a child, a woman or couple offers, in addition to a fee, an implied promise of gratitude for what is often considered by both parties to be a priceless gift. Still, despite retaining the gift rhetoric of its early years, egg donation was already well on the way to becoming a commercial operation before the advent of frozen eggs. Using an egg donor from an established agency can easily cost $37,000 or more, according to ConceiveAbilities, a donor egg brokerage with offices in metropolitan areas on both coasts and in Chicago. In addition to donor compensation, which at this agency starts at $7,000, added charges include medical expenses (the intended mother's and the donor's), agency fees, and other costs.[44] Will frozen eggs purchased from an online catalogue appreciably lower those costs? Probably not. Eight of "Ariana's" eggs cost $20,000, which did not include the costs of fertilization and embryo transfer procedures.

Some agencies have begun to offer frozen eggs for online purchase in addition to their regular services. It seems to us, however, that the option of ordering frozen eggs online the same way people purchase consumer goods may undercut the older model of egg donation, making the process more impersonal.[45] The potential impact of the online egg market is not easy to predict. Even before its advent there were more patients seeking to conceive with donor eggs, and in 2015, 91 percent of fertility centers offered services to individuals and couples wishing to use them. It is true that the percentage of donor egg users has declined a tiny bit—from about 13 percent of retrieval cycles in 1999 to just under 12 percent in 2015—but the absolute numbers have increased. In 1999, there were about 9,000 cycles using donor eggs or embryos (in those days, the two were not differentiated). In 2015, donor eggs were used in 19,482 cycles.[46]

Donating Embryos—or Selling Them?

Donor eggs and donor sperm are the most common ways in which a third or sometimes a fourth person becomes part of one of the most intimate experiences of a couple's life—the conception, gestation, and birth of the next generation of a family. The other two are embryo donation and gestational surrogacy. Both are rare. Embryo donation is generally little discussed outside the profession, and surrogacy, in contrast, has garnered such outsized attention that many people believe it to be a much more prevalent phenomenon than it is.

We just used the term "embryo donation," but there are several ways by which an intended mother might find herself giving birth to a baby to whom she, or she and her husband or partner, are unrelated genetically. One form of embryo donation, as we noted in chapter 6, is truly altruistic. Some couples, once they have successfully conceived, tell the reproductive specialist who treated them that their extra embryos can be offered to another patient.[47] But most couples choose not to do that. Many couples simply cannot decide what to do; indecision is the likely reason for the vast numbers of embryos remaining in storage for years, even decades. In 2002, at the time of the most recent attempt to count those stored embryos, there were almost four hundred thousand frozen embryos in facilities across the nation. By now, according to some estimates, there could be a million or more.[48]

One California doctor decided that since existing frozen embryos were not going to be available in any significant numbers, he would simply create new ones. Ernest Zeringue began marketing ready-made embryos created from donor eggs and donor sperm in 2010, shocking his colleagues. Even outspoken advocates of the use of donor gametes and gestational surrogates believed he had crossed a line.[49] But could anything be done about his actions?

Several reproductive endocrinologists and ethicists weighed in on the problem. Now that "the prospect of a for-profit embryo bank [was] no longer theoretical," said distinguished reproductive endocrinologist Eli Adashi and Harvard law school professor Glenn Cohen in an article about Zeringue's program, it was time for the field of reproductive medicine to examine both the ethics and potential consequences of creating embryos with the specific intent of selling them. If the sale of sperm and eggs individually was considered ethically acceptable, they said, should we consider the sale of embryos to be the same or different? In the end, they concluded, the ethical issues were no different whether the gametes were offered separately or already combined. Nevertheless, Adashi and Cohen recommended that the American Society for Reproductive Medicine (ASRM) develop ethical guidelines to govern the practice. They also urged states to provide "clear legal guidance" regarding the parenthood of the children who would be born from these embryos.[50] Columbia University's Mark Sauer, however, disagreed, calling the deliberate creation of embryos for sale more ethically fraught than the sale of eggs or sperm separately, because an embryo can become a human being. Even if there were no other ethical issues involved, he and psychiatrist Robert Klitzman argued, that fact alone "should prompt us to revisit, rather than ignore these ongoing controversies in this new context."[51]

Adashi, Cohen, Sauer, and Klitzman put their faith in professional guidelines to enforce standards and uphold ethical norms, recommending that fertility centers offering ready-made embryos report this practice to the ASRM, limit the numbers of people allowed to purchase embryos from the same donors, and pledge not to charge higher prices for the embryos of donors possessing nonmedical traits such as beauty, intelligence, or height. They put their confidence in the willingness of physicians to follow such guidelines and assumed that patients would not pressure their doctors to go beyond them. Was that possible? After all, eggs and sperm from donors considered to have such sought-after nonmedical traits already commanded higher prices, so exactly how the ASRM could persuade embryo sellers to behave differently from sperm banks, egg donor agencies, and egg banks is difficult to imagine. Nevertheless, although specific guidelines have yet to materialize, the overt disapproval from leaders in the field may have had an effect. To date, California Conceptions, Zeringue's operation, does not seem to have been imitated or replicated elsewhere.[52]

California Conceptions calls itself a "donor embryo" program, even as it engages in market transactions. It buys (eggs and sperm) and it sells (ready-made embryos). Still, "donor" sounds better than "buyer and seller." So does the term "adoption," at least to one evangelical Christian agency that earns a fee for providing infertile couples with the leftover embryos of fellow evangelicals who have agreed to donate them. The first "embryo adoption" program, called Snowflakes, was inaugurated in 1997. A branch of Nightlight Christian Adoptions, Snowflakes has specific requirements for couples who wish to utilize its services. Embryos are available only to like-minded Christians, who must undergo a home study to prove their religious bona fides. President George W. Bush, who supported the organization's anti-abortion mission, provided federal funding of about $1 million to Snowflakes in the early twenty-first century and gave the agency a public relations boost by posing with a group of its babies. In the twenty years since its founding, more than 550 babies have been born through the program. Acquiring a Snowflakes embryo is not inexpensive. With an agency fee of $8,000, having a baby can cost future parents from around $13,500 to $18,000, or even more, and that's if they achieve a pregnancy on the first try. The cost is higher if it takes longer than one cycle to become pregnant.[53]

The other leading anti-abortion "embryo adoption" program is run by the National Embryo Donation Center (NEDC), founded in the early twenty-first century in Knoxville, Tennessee. The center's website declares that it was founded on a belief "in the sanctity of life beginning at conception" and has a mission "to protect the lives and dignity of human embryos." Sponsored by the Christian Medical and Dental Association, the center has been directed since its inception by Jeffrey Keenan, a local reproductive endocrinologist. Unlike Snowflakes, which brokers transactions between couples who have excess embryos and those who want to use them, the NEDC is a medical facility. Embryos, donated altruistically, become the property of the center, which stores them on site. It does not, however, turn around and donate them to others at no cost. The NEDC, like Snowflakes, has policies to govern what kinds of couples can use those embryos—no single women or same-sex couples are eligible, and couples who seek embryos must share the NEDC's religious beliefs and behave accordingly. Like Nightlight, Snowflakes' parent organization, the NEDC insists on a home study to ensure that embryos will only go to those who embody its values. The costs of that home study, plus agency fees, are added to any costs that couples incur for the embryo transfers. It appears that Keenan performs those transfers at the fertility center he directs in Knoxville. In 2015, his center initiated 185 IVF cycles, 75 percent of which involved frozen embryos, presumably from the NEDC. In 2017, the baby counter on the NEDC's website showed that 686 babies had been born since the center began operation in 2003.[54]

The ASRM rightly disapproves of the misleading and politically charged term "adoption" to describe the process of the use of donated embryos. An embryo cannot be adopted because embryos are not persons and the laws of the United States do not recognize them as such. An embryo might become a baby, of course, but there is no guarantee that it will. It may not implant. If it does implant, it may be miscarried. Calling the use of another couple's embryos "adoption" is a political decision made by organizations who oppose abortion.

However this process of becoming pregnant using the embryo of someone else is proposed—accept an altruistic donation from a couple who has completed their family, purchase a ready-made embryo, or "adopt" one from Snowflakes or the NEDC—the practice remains rare. The only real measure we have of the interest of couples in using someone else's embryo to conceive their child is the national clinic data. In 2015, out of 232,000 retrieval cycles, just 1,700 of them involved donor embryos, compared to more than 19,000 using donor eggs.[55]

Bearing a Child for Someone Else:
The Complexities of Gestational Surrogacy

"I did not give birth to my son," wrote Alex Kuczynski in the *New York Times Magazine*. At the age of thirty-nine, after five years of trying to conceive, eleven failed cycles of IVF, and four miscarriages, she and her fifty-four-year-old husband decided to "give gestational surrogacy—hiring a woman to bear our child—one try."[56] With embryos created from Kuczynski's eggs and her husband's sperm, surrogate Cathy Hilling gave birth to their son in 2008. Kuczynski was ready to share her story with the world. She was surely not, however, prepared for the firestorm of criticism it produced. Beginning with the cover photo, featuring a slim, expensively dressed, and beautifully made-up Kuczynski standing next to a very pregnant Hilling wearing wrinkled khakis and a button-down overshirt, and moving on to Kuszynski's surprise that Hilling owned a computer and knew "how to use it," the story seemed to exude class privilege and condescension. Or so concluded most of the readers who wrote to the *New York Times* in response to it. Many of those who commented on the story were particularly struck by the article's photographs. "The surrogate mother is sitting, barefoot, on a dilapidated porch in one photo," one of them observed, "whereas the mother and child are standing in front of a hugely expensive, well manicured home with their baby's nurse, a black woman, in the other photo."[57]

Hilling, the surrogate, said in response to these comments that perhaps the *Times* editors or the photographer wanted to make a point that Kuczynski herself hadn't intended. The photographer staged the photos, Hilling said, asking her to take her shoes off for the one on the porch. "There were lots of photos taken in my beautiful yard. . . . But those photos would not have gotten a rise out of the readers." It wasn't just the photographs, however. The entire tone of this piece, about a wealthy couple who could easily afford the $100,000 or more to hire someone else to bear their child, radiated a sense of entitlement, stunningly revealed by the author's seemingly offhand admission that she hadn't expected these middle-class people who lived in Harleysville, Pennsylvania, to be so intelligent and informed. Maybe, though, Kuczynski was just being honest when she explicitly portrayed her relationship with her surrogate as transactional. Of course, we might say, surrogacy *is* a transaction: Cathy Hilling was being paid to bear another couple's genetic child. But readers wanted Kuczynski to show some understanding of the discomfort and perhaps even peril that a pregnant woman faces, and to see Hilling as so much more than a rented uterus. Instead, it was Hilling who made the gestures of kindness that can make the process of surrogacy more than just a commercial exchange. "When the baby was born," Kuczynski wrote, "she was the one who thought to bring a gift for me to the hospital."[58]

There were other intended parents, in contrast, who expressed both gratitude and empathy for their surrogates. Eric Ethington and Doug Okun forged a bond of mutual respect with their surrogate, Ann, and her husband. Eric and Doug had known soon after they met in the late 1990s that they wanted children. "From literally our second date, I was asking Doug how he felt about having children and getting married, because I wanted both those things so much," Eric told a reporter from the *San Francisco Chronicle* in 2004. "I told him these are deal-breakers. Luckily, those things were important to him, too." Their commitment ceremony, the only substitute for marriage available to same-sex couples before they gained the legal right to marry, was held in 2000 in the Napa Valley. "About 100 people, a lot of kids," said Doug. "It was really one of the most beautiful days of my life."[59] Settled down, the couple were ready to start their family and turned to Growing Generations, a Los Angeles surrogacy agency that specializes in creating families for same-sex couples. By 2003, they had found their surrogate, selected an egg donor, and were awaiting the birth of twin girls. Half of their embryos had been conceived using Doug's sperm, the other half with Eric's. Their twin daughters, one of them Eric's genetic daughter and the other Doug's, were born on November 7. Four months later, the babies and their fathers were on the front pages of newspapers across the country when Doug and Eric were married at San Francisco's City Hall soon after the city's mayor announced that he was authorizing same-sex marriage in the city.[60]

Gestational surrogacy is the most expensive form of assisted reproduction, costing anywhere between $80,000 and $150,000, depending on a variety of factors, including whether the egg of the intended mother will be used or that of a donor, the individual fee for the surrogate, and agency fees. Doug and Eric were both successful financial professionals and could afford the cost.[61] What author Liza Mundy said about surrogacy—that having a baby this way means "entering a world that combines profound parental love with cold-blooded business truths, a world where children are desired, loved, celebrated, wanted, and obtained in a relentlessly commercialized process"[62]— is true of every reproductive technology in which gametes are bought and sold. But as we can see from the dismayed comments, from so many readers, that greeted Alex Kuczynski's story, compensated gestational surrogacy evokes stronger feelings than egg or sperm donation because it involves paying a woman to bear a child and to risk the potential medical hazards that a pregnancy may involve. These risks can be substantial. Pregnancy is often without incident, of course, but it can sometimes entail life-threatening complications. Ann, Doug and Eric's surrogate, nearly died from a postpartum hemorrhage.[63] Wanda faced this very situation when she was a practicing obstetrician. A surrogate, carrying a baby for a woman who had had a hysterectomy for cancer, suffered a severe postpartum hemorrhage. A mother of three, she was pregnant for the fourth time as a surrogate. Her pregnancy was uneventful, and she had developed a strong emotional bond with the genetic mother of the newborn she delivered. During that delivery she almost died; what began as a joyous birth nearly ended in tragedy.

Even altruistic surrogacy is not without its ethical hazards. Frana and Susan grew up in the same Texas town, where their fathers had been best friends. Susan had a congenital heart condition. The drug used to treat it was known to cause severe birth defects, which meant that it could not be taken during pregnancy. When her cardiologist told her that it was "absolutely unsafe for her to go off the drug for nine months of pregnancy," she tried to come to terms with the devastating news that she would never have children and that her husband, Scott, would never be a father. When Susan's mother told Frana how unhappy Susan was, Frana immediately said, "I'll have a baby for her."[64] She had two of Susan and Scott's embryos transferred into her uterus. They were soon shocked to learn that she was carrying triplets. Decisions had to be made: Would Frana undergo a selective reduction to one or two fetuses, terminate the pregnancy entirely, or stay the course and hope for the best? She was pregnant for a close friend, but she was running significant health risks, and "from a legal perspective, the choice of whether to proceed with the pregnancy" was her decision. She knew that carrying triplets "increased her chances of severe complications such as preeclampsia, diabetes, and preterm labor," but she also worried about disappointing Susan. In the end, Frana continued with the pregnancy, and Susan decided to rent a condo next door to Frana and her family, taking over Frana's household duties and child care responsibilities until after the babies were born. Frana was thirty-four weeks pregnant when, according to her doctor, "her kidney and liver functions were declining." He delivered her by C-section. The babies, the article reported, were small but healthy, and Frana had no regrets. Everyone involved saw it as a happy ending.

Other stories were chilling. Heather Rice had three children of her own when she became a surrogate for the first time. The experience, she said, was "so great I knew I wanted to do it again." The second time, however, the situation turned ugly. A routine ultrasound at twenty-one weeks showed that the fetus had "a cleft in his brain." The medical name for this condition is schizencephaly, a rare genetic disorder that can cause seizures, paralysis, and developmental delays. Depending on multiple factors, schizencephaly can range from mild to severe.[65] When it became evident during the ultrasound that there was a problem, Heather was shocked that "Mom walked out of the room, left me lying there, and I thought: 'This is not my baby. I should not be dealing with this by myself.'" After the scan, the genetic parents informed Heather that they would not take the baby and told her to have an abortion. It was so late in the pregnancy that she found herself unable to do so, taking it upon herself to locate adoptive parents for the baby, a couple who already had one child with the condition. She told the parents, who responded by requesting a second scan. They then told Heather they had decided to keep the baby. Heather found it strange that only the father showed up at the birth, and he took the newborn away shortly thereafter. Becoming ever more anxious about the baby's fate, she searched for and found the couple on Facebook. There "was no trace" of a baby, "so I think they gave him up for adoption," she told a reporter for the *New York Times*. "I don't know where he is, and it kills me every day."[66]

The media pays a great deal of attention to surrogacy, and sometimes stories include high but unverifiable numbers of babies born using the procedure. This can leave the impression that surrogacy accounts for significant numbers of the children conceived using assisted reproduction. But in reality—perhaps because of the expense, or the potential health risks to surrogates, or discomfort over the idea of another woman bearing a child for someone else, or some combination—surrogacy remains a little-utilized reproductive option. In 2015, about 1 percent of egg retrieval cycles—about 2,300—were performed at fertility clinics in the United States for gestational surrogates.[67] A recent study of gestational surrogacy from 1999 to 2013, which uses a different unit of analysis—embryo transfers rather than egg retrievals—showed that gestational carriers were involved in just under 2 percent of embryo transfers over that period. In 2013, just under 2.5 percent of embryo transfers involved surrogates.[68]

Surrogates, who are mostly young and fertile, have a higher rate of fertilization, pregnancy, and births than do women who are undergoing IVF because of their own or their male partner's infertility, so they need fewer cycles overall to achieve a pregnancy. This is true even though gestational surrogates do not use their own eggs; about 50 percent of them use the egg of the intended mother, and the other 50 percent are using a donor egg.[69] In either case, they have a higher rate of successful embryo transfer, pregnancy, and birth than do infertile women using their own eggs. For the period between 2009 and 2013, about 2.2 percent of deliveries after IVF were to gestational carriers, and for the year 2015, it was slightly above 3 percent.[70]

Over time, it has become much easier for those who want to use gestational surrogacy to access treatment. Wulf Utian's clinic at Mt. Sinai Hospital in Cleveland may have been the only one offering the service in the mid-1980s, but by the end of the decade, 46 percent of reporting clinics indicated a willingness to treat gestational carriers. By 2015, a total of 87 percent did so.[71] There has also been an influx of foreign nationals seeking to use gestational carriers in the United States because most other countries forbid paid surrogacy. In 2014, the *New York Times* reported that affluent Europeans, Asians, Australians—"gay, straight, married, or single"—were coming to the United States for surrogacy services.[72] In 2013, international patients accounted for about 18 percent of embryo transfers in American fertility centers.

In 2015, the US Supreme Court ruled, in a 5–4 decision, that same-sex couples have a constitutional right to marry. As Justice Anthony Kennedy wrote for the majority, "No union is more profound than marriage, for it embodies the highest ideals of love, fidelity, devotion, sacrifice and family. In forming a marital union, two people become something greater than once they were."[73] Same-sex marriage had been legal in some states for over a decade by the time the matter came before the Supreme Court, but state-by-state rulings do not have the force of national law.

The ruling's impact was immediate. According to one study, four months after the court's decision, 45 percent of same-sex couples were married, up from about 21 percent two years earlier. In addition, about 25 percent of same-sex couples, married and unmarried, were raising children. It is impossible to know what impact marriage equality will have on the use of gestational surrogacy by male couples. Among those using gestational surrogates in the twenty-first century thus far, about 11 percent have been same-sex male couples.[74] It is not certain that greater numbers of marriages among male couples will lead to higher utilization of surrogacy. Today, many of the children being raised by same-sex couples are adopted, and same-sex couples already adopt at a rate 4.5 times that of heterosexual couples. It is possible that marriage equality will lead to both higher adoption rates and greater use of medicine and technology by same-sex couples generally, but it is too early to tell.[75]

Egg Freezing—Promise or Peril?

"It's not unusual for a single woman in Manhattan to shell out $45 at a boutique hotel for a few cocktails and paltry appetizers," wrote Amanda Duberman in the *Huffington Post* in 2014. "It's a little less common to leave said hotel with a $1,000 coupon to freeze her eggs." Duberman had just been to a cocktail party called "Let's Chill" hosted by Eggbanxx, a for-profit egg-freezing startup. The organizers had expected forty people. About a hundred showed up, most of them women "in their early 30s."[76] Could egg freezing really offer the promise of future fertility? Sarah Elizabeth Richards surely hoped so when she froze her eggs in her mid-thirties, spending $50,000 to do so and wiping "out her savings and the money her parents had set aside for [her] wedding." It was worth it, Richards said: "the best investment I ever made." Richards turned her experience into a book in which she followed the lives of four other women who froze their eggs. Only one of them ended up having a baby from these eggs. Richards appears either never to have used hers, or perhaps they failed to work. Readers of her book *should* have seen the moral of the story as a warning: Spend tens of thousands of dollars to freeze your eggs, and you may easily end up poor and infertile. That was not, however, the lesson Richards took away from the experience nor the one she wanted to impart. To her, these women's experiences in egg freezing represented their empowerment. "Their faith that egg freezing would work set in motion positive events in their lives," she wrote. "They enjoyed years with less baby panic, comforted that they were getting a second chance at motherhood."[77]

Doctors who began trying to accomplish oocyte cryopreservation (the medical term for freezing eggs) in the 1980s had other things on their minds besides cocktail parties where fertile young women were urged to freeze their eggs. The world's first birth from a frozen egg occurred in 1986. The technology then in use was unreliable, however. Few frozen eggs, once thawed, were capable of being fertilized. But in 1997, Italian researchers reported achieving a pregnancy from eggs cryopreserved using a new method called vitrification, a faster freezing technique that avoids the problem of ice crystal formation.[78] It was a breakthrough. By the early twenty-first century, it was clear that eggs cryopreserved with this new technique were much more likely to be fertilized after thawing than those frozen by the older technique.

This new method was especially welcomed by doctors treating young women with cancer who were about to undergo treatments that could destroy their reproductive capacities.[79] Even as they sought a cure for a life-threatening disease, these patients did not want to forgo all hope of ever having children. If she was married, a patient might have her eggs retrieved and fertilized with her husband's sperm, freezing the resulting embryos. But what about women without a partner, or adolescent girls? Now they could have their eggs frozen with at least some hope that those eggs might be viable in the future. At this time, oocyte cryopreservation was considered an experimental procedure, and the ASRM recommended that it be used only for medical indications. And although recommendations are just that, recommendations, the profession seems to have heeded them. In 2009, fewer than five hundred women froze their eggs. Most of them were probably cancer patients. Then, in 2012, the Society for Assisted Reproductive Technology (SART) completed a review of data on nearly a thousand births from frozen eggs, discovering that pregnancy rates were similar to those from fresh eggs and saw no increase in birth defects in children born from frozen eggs as compared to fresh ones. As a result of these findings, the ASRM then removed the procedure's experimental designation.[80]

Although the ASRM had declared egg freezing to be safe, it did not encourage members to offer it electively, on the grounds that "marketing this technology for the purpose of deferring childbearing may give women false hope."[81] The society's cautionary words went unheeded. For-profit companies, in particular, saw an opportunity to take advantage of the worries among thirty-something women that their fertile years were slipping away. The founder of Extend Fertility, Harvard Business School graduate student and entrepreneur Christy Jones, had no medical background and was unapologetic about seeking a way to profit from women's anxieties. She hired public relations specialists, built a website, and engaged in direct marketing efforts with such tags as "Set your own biological clock."[82] The idea that a woman could freeze her eggs and postpone pregnancy until she was ready for motherhood, she believed, would appeal to women who might someday want to become pregnant but were not ready, not married, not sure, or all three.

Before too long, established IVF centers also began offering elective egg freezing. In 2013, just a year after the lifting of the experimental label, nearly four thousand women froze their eggs. These eggs are not guaranteed to fertilize, however. A woman who freezes her eggs at the age of twenty-five has about a 28 to 31 percent chance of having a baby later with six of those thawed eggs. If she waits until she is thirty-five to freeze them, her chances drop to about 16 to 18 percent. For a woman of forty, they are 12 to 13 percent. And those chances, remember, are for six eggs, not one or two or three. Even for young women, egg freezing offers no promises. The ASRM's current website warns women in stark terms not to have high expectations of their frozen eggs. "Even in younger women [who freeze their eggs under the age of thirty-eight], the chance that one frozen egg will yield a baby in the future is around 2–12%." Those who are older than that have even less of a chance.[83]

Women who freeze their eggs apparently do not want to hear these statistics. If Sarah Richards is at all representative, they believe that freezing their eggs will allow them to postpone motherhood as long as they choose. Richards was outraged that the ASRM wanted physicians to tell their "customers every possible reason they shouldn't buy their service" before they would agree to freeze a woman's eggs. She thought that the organization was "cherry-picking the most discouraging statistics" to keep all but the youngest women from believing egg freezing would work for them.[84] She and other women in their mid to late thirties were not to be deterred from paying $10,000 to $15,000 per cycle, plus additional fees for storage, to freeze their eggs. But are they entirely responsible for their misplaced beliefs? Aggressive advertising by for-profit companies make exaggerated claims, and those ads are difficult to escape. As one young woman told us, egg-freezing ads pop up on her laptop with disconcerting regularity. Businesses with names such as Extend Fertility and Eggbanxx, just to name two of the more ubiquitous ones, host cocktail parties to promote their services, and fertility centers across the country offer the procedure.

According to a 2017 survey, 81 percent of SART member clinics offer egg freezing, and 88 percent of those clinics offer elective egg freezing—75 percent of academic centers and 91 percent of private clinics. The other 12 percent provide the service only for medical reasons. Few of them include information on their websites regarding cost or effectiveness. Just 17 percent of the clinics mentioned cost, and only 10 percent cautioned "that [oocyte cryopreservation] is not a guarantee [of] future fertility."[85] If more of these clinics told women directly that egg freezing is expensive, uncomfortable, not risk-free, and comes with no promise of future fertility, would it even make a difference? It doesn't matter. It's their obligation to provide women with the facts.

Fertile Couples and Assisted Reproductive Technology

An infertility specialist with whom we are acquainted, now in his seventies, spent many years in academic medicine and trained numerous residents and fellows. A few years ago, he visited one of them, a prominent practitioner whose IVF practice included a significant number of wealthy couples from parts of the world where sons are more highly valued than daughters. These couples seek his services not because they can't conceive on their own, but to undergo IVF with preimplantation genetic screening (PGS) to ensure that their baby will be a boy. When the young doctor proudly said that his clinic had very high success rates, the older physician said he couldn't help but sigh. "Of course he has high success rates. These patients aren't infertile."[86]

In its specifics, this is a story about preferring boys to girls and having the money to choose a baby of the desired sex. More broadly, it is just one example of the use of assisted reproductive technologies for purposes that go beyond infertility. Such purposes have a longer history than some might suspect. In the United States, the IVF pioneers of the early 1980s were focused on using the technology as a treatment for infertility. But it was different in England, where, beginning with his early work, Robert Edwards argued that IVF might help reduce and perhaps eventually eliminate the occurrence of heritable diseases, if only they could be diagnosed in the embryo. This was one of the key arguments made in Great Britain by medical scientists in their quest to legitimize the new technology. In the end, public and governmental support for IVF in Great Britain in the 1980s was driven in some considerable measure by its promise to conduct the research needed to allow women to have babies without genetic disorders. Fulfilling Edwards's vision, preimplantation genetic diagnosis (PGD) was first developed in England in the late 1980s and early 1990s.

PGD is a diagnostic test that can determine whether an embryo contains a specific disease-causing gene. Couples in which one or both partners are known or suspected carriers of a heritable disease can undergo IVF and then have their embryos tested for the disease-causing gene. Examples of such inherited disorders include cystic fibrosis, Huntington disease, and sickle cell anemia. In 1992, British biologist Alan Handyside and four colleagues tested the embryos of a couple who were both carriers of cystic fibrosis with the intent of ensuring that the embryos to be transferred did not have the disease.[87] The Human Fertilisation and Embryology Authority (HFEA), which regulates assisted reproduction in Great Britain, lists a total of four hundred diseases for which PGD has been approved.[88]

Preimplantation genetic screening is a different type of test in which embryos created by IVF are screened for chromosome abnormalities or sex before being transferred into the uterus. Humans have twenty-three pairs of chromosomes, for a total of forty-six. PGS is not used to *diagnose* a specific disease; it provides information on the number of chromosomes. Embryos without aneuploidy (an abnormal number of chromosomes) can then be implanted. Down syndrome (also known as trisomy 21) is an example of aneuploidy—the embryo has an extra copy of chromosome 21.[89]

Another technology, called mitochondrial replacement techniques, or MRT, makes it possible to create embryos free from diseases that are transmitted via the mother's mitochondria. Nearly every cell contains mitochondria, which are organelles that produce energy. Humans have two kinds of DNA, and they have different functions. Nuclear DNA (nDNA) gives us most of our heritable features, such as hair and eye color. When we say, "He looks just like his mother," we are talking about inheritance of traits through nDNA. Mitochondrial DNA (mtDNA) contains very few genes; these genes encode vital components of energy production. With MRT, the nuclear DNA is removed from the mother's unfertilized or fertilized egg and then transferred to the egg of a donor who has normal mitochondrial DNA. This allows a woman with a disease due to abnormal mtDNA to bear children who will have her nuclear DNA but not inherit her defective mitochondria. Because these techniques require an egg donor, there are three separate people contributing genetic material to the subsequent baby—the intended mother, the father, and the donor.

The idea of using a donor's mtDNA in IVF is controversial. The United States was not the only country to ban the use of donor mitochondria. Great Britain did, too. In 2015, however, the British Parliament reversed course, and mitochondrial replacement techniques, although strictly regulated and limited in use, are now legal there.[90] At the same time, in the United States, the Food and Drug Administration began to rethink its earlier prohibition. It asked the Institute of Medicine (now the National Academy of Medicine) to study the ethics of using MRT to prevent the transmission of mitochondrial diseases from mother to child. This would allow women who carry mitochondrial disease to bear their own biological children.

The FDA's authority over MRT does not derive from the ban on federal funding for embryo research. Rather, the agency considers these techniques a form of gene therapy, which it already regulates, because they modify human eggs or embryos. Genetic modification is the key here: these techniques permanently change the mitochondrial DNA of the child. And if the child were a girl, she'd pass down those modifications to future generations. The Institute of Medicine created a committee to study the issue, which concluded, after months of study, that the use of MRT, under these specific circumstances, was ethical. It also recommended limiting any proposed embryo transfers to male embryos until the long-term effects on future generations could be understood. Why only boys? Any changes to a male child's mtDNA would not be transmitted to the next generation. Since mtDNA is not present in sperm, only female children could pass the donor mtDNA on to subsequent generations. Allowing only male embryos to be considered for transfer was designed "to avoid," in the committee's words, "introducing heritable genetic modification during initial clinical investigations."[91]

Mitochondrial replacement is controversial for several reasons. In female babies, it alters the genetic make-up in ways that may not yet be understood. Some ethicists are troubled by the possibility of unforeseen long-term consequences. There are other concerns, as well. Disability rights advocates have argued that the use of MRT could devalue the lives of people with disabilities. Abortion opponents object on the grounds that research in this area would invariably destroy some human embryos. Lastly, the familiar "slippery slope" argument has been invoked. Once mitochondrial replacement is allowed for the prevention of disease, this contention goes, before too long it will be used electively, perhaps in older women whose eggs no longer can be fertilized, perhaps for as yet unknown reasons.[92]

In the end, however, the Institute of Medicine's recommendations were unheeded. About six months before they were released, Republicans in the House of Representatives had inserted a provision into the appropriations bill that explicitly prohibited the FDA from using federal funds to approve any research on MRT. "None of the funds made available by this Act," the language read, "may be used to review or approve an application for . . . research in which a human embryo is intentionally created or modified to include a heritable genetic modification." As a result, within a few hours after the Institute of Medicine panel issued its recommendations, the FDA announced that such research "cannot be performed in the United States" at this time.[93]

Between this provision of the appropriations act and the more than two-decade-old Dickey-Wicker amendment, which prohibits the use of federal funds for any research in which a human embryo is either created for research purposes or destroyed as part of the research, the FDA had no clear path forward. The 2016 appropriations act prohibited the FDA from using federal funds to accept investigational new drug applications—called INDs—that sought to conduct research studies on human MRT. So even though Dickey-Wicker has no power to stop *private* research, the two laws together created a one-two punch. One, the FDA has said that MRT can't go forward without its approval. Two, the 2016 federal law *prohibits* the FDA from reviewing proposals for research. The end result? It doesn't matter whether the private sector would be funding the research—the FDA gets the final say over whether research can proceed. But because the FDA is not allowed to use federal money to *approve* any INDs, its hands are tied. Unless Congress eliminates these provisions, hardly likely in the current political climate, research and clinical applications of mitochondrial replacement techniques will not occur in the United States.

Congress's handling of the careful efforts of the FDA and the Institute of Medicine to allow research on MRT is characteristic of its approach to the new reproductive technologies since their use was first being contemplated in the mid-1970s. This appropriations rider, like the Dickey-Wicker amendment, emanated from members of Congress opposed to abortion. The end result was to shut down the FDA's efforts to support research on MRT. But if it was able to shut down regulated research, it could not keep those determined to conduct such research from evading the law by performing the procedure in another country. John Zhang, the director of New Hope Fertility Center in New York, employed MRT to enable a woman, whose child would have otherwise been afflicted with a fatal disease called Leigh Syndrome, to have a baby boy free of that disease. To evade the FDA's ban, Zhang created the embryos in New York but transferred them into her uterus in Mexico. After a stern warning from the FDA in 2017, he has apparently agreed not to create any more embryos in the United States, but that will not prevent him from doing so in Mexico.[94] The FDA's attempts to create a scientifically sound and ethically defensible process for dealing with this issue failed, and as a result, mitochondrial replacement techniques will be developed outside the confines of federal regulation and available only to those who can pay at clinics beyond our borders. It is one more chapter in a long history of failed efforts. In our next and final chapter, we seek to make sense of this history as we ask whether the Wild West of reproductive medicine might yet be tamed.

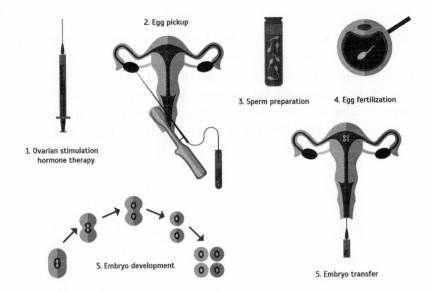

2. Egg pickup

3. Sperm preparation   4. Egg fertilization

1. Ovarian stimulation
hormone therapy

5. Embryo development

5. Embryo transfer

Figure 11. This diagram provides a simplified schematic of the in vitro fertilization process. First, the patient receives medication to stimulate her ovaries so that they will produce multiple eggs. The next step is the egg retrieval process, followed by fertilization with sperm provided by a male partner or donor. If the fertilization is successful, the resultant embryos are transferred into the uterus. Today, embryos are often frozen for a period of time before transfer. From iStockphoto.com.

Figure 12. In the early days of in vitro fertilization, many of the doctors who performed the procedure cared for their patients throughout pregnancy and even beyond. Stephen Sondheimer was one of them. He is shown here with his patient Wendy and her twins Justin and Brent at the babies' bris in 1984. Wendy's babies were the first twins born as a result of IVF in Philadelphia. Used with permission of the patient.

Figure 13. Alan DeCherney (*left*) and Luigi Mastroianni, 1988. Mastroianni, who had been mentored by John Rock, was in turn a mentor to DeCherney, a legacy of IVF stretching across generations. Courtesy of Dr. Christos Coutifaris and the Division of Reproductive Endocrinology and Infertility, Perelman School of Medicine, University of Pennsylvania.

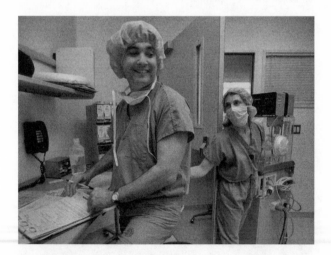

Figure 14. Richard Marrs and Jody Greene in their laboratory in 1997. Fifteen years earlier, in 1982, the tiny in vitro fertilization program run by Marrs at the University of Southern California was the second in the United States to achieve a live birth using in vitro fertilization. Jody had been his technician for that first birth in 1982, and as of 2018 she was still working with him. Photo by Ken Hively. Copyright 1997, *Los Angeles Times*. Used with permission.

"I'm their real child, and you're just a frozen embryo thingy they bought from some laboratory."

Figure 15. The first American birth from a frozen embryo occurred in 1986, but as the cartoon suggests, some discomfort with the technology was still evident in the late 1990s, especially when embryos frozen at the same time produced babies several years apart. *New Yorker*, January 19, 1998. William Hamilton / The New Yorker Collection / Cartoon Bank. Used with permission.

Figure 16. Ricardo Asch in 1996. Asch, one of the most prominent infertility specialists in the United States, fled the country in 1995 after being accused of stealing eggs and embryos from patients and then using them in other patients. He remained a fugitive from justice for more than fifteen years before his arrest in Mexico, which refused to extradite him. Photo by David Fitzgerald. Copyright 1996, *Los Angeles Times*. Used with permission.

"Arthritis Pain Formula Anacin, Poli-grip, and a home-pregnancy test."

Figure 17. In the 1990s, some IVF programs were taking women up to the age of fifty-five, but the practice was controversial. *New Yorker*, January 17, 1994. Edward Frascino / The New Yorker Collection / Cartoon Bank. Used with permission.

Figure 18. Eric Ethington and Doug Okun with their twin daughters on February 13, 2004, at their wedding at San Francisco's City Hall. Photo by Paul Chinn for the *San Francisco Chronicle* / Polaris Images. Used with permission.

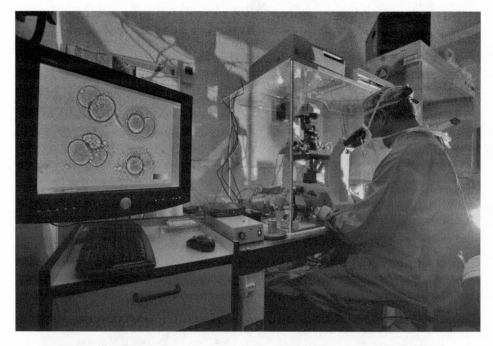

Figure 19. IVF Laboratory, 2008, with images of embryos on the computer screen. This photograph appeared for the first time in October of 2008, accompanying an article about embryos left in storage after IVF and how former patients thought about what to do with them. It was used again in 2010 for an article about preimplantation genetic diagnosis. The physician pictured is David Diaz, an infertility specialist in the Los Angeles region. Photo by Mark Boster. Copyright 2008, *Los Angeles Times*. Used with permission.

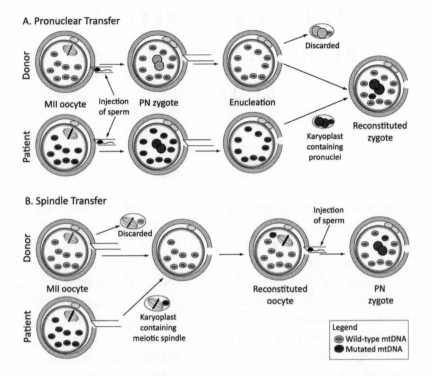

A. Pronuclear Transfer

Donor

MII oocyte — Injection of sperm — PN zygote — Enucleation — Discarded

Patient

Karyoplast containing pronuclei

Reconstituted zygote

B. Spindle Transfer

Donor

MII oocyte — Discarded — Reconstituted oocyte — Injection of sperm — PN zygote

Patient

Karyoplast containing meiotic spindle

Legend
- Wild-type mtDNA
- Mutated mtDNA

Figure 20. Two mitochondrial replacement techniques. In pronuclear transfer (*A*), both the egg of the intended mother and the egg of a donor are fertilized, the nucleus of each fertilized egg is removed, and the defective mitochondria of the intended mother's embryo is replaced by the healthy mitochondria of the donor. In spindle transfer (*B*), the part of the egg containing the intended mother's defective mitochondrial DNA is removed prior to fertilization, as is the part of the donor's egg containing the nuclear DNA. The intended mother's nuclear DNA is then inserted into the donor egg, after which it is fertilized. One of the differences between the two techniques is that in pronuclear transfer, two embryos are created, and in spindle transfer, just one. Creative Commons Attribution License. Jessica Richardson, Laura Irving, Louise A. Hyslop et al., "Concise Reviews: Assisted Reproductive Technologies to Prevent Transmission of Mitochondrial DNA Disease," *Stem Cells*, February 17, 2015.

8

CAN THE WILD WEST OF
REPRODUCTIVE MEDICINE
BE TAMED?

In 2011, a Beverly Hills fertility specialist lost his medical license for gross negligence after he transferred twelve embryos into a thirty-two-year-old patient who subsequently gave birth to eight babies.[1] The doctor was Michael Kamrava, and his patient, Nadya Suleman, soon became internationally notorious as the "Octomom." When her octuplets were born in 2009, Suleman had six other children, also conceived by IVF at Kamrava's fertility center. She was unemployed. She and the children lived with her parents. Kamrava later claimed that he had initially balked at implanting so many embryos, agreeing only because Suleman insisted. That excuse rang hollow. It is impossible to imagine any responsible reproductive endocrinologist implanting twelve embryos into a young and demonstrably fertile patient.[2] And this was not Kamrava's only offense. The California Medical Licensing Board found gross negligence in his treatment of two other patients. Beyond malpractice, he was known for transferring a larger number of embryos into young patients than were recommended, and anyone who looked up his IVF success rates could see that they were significantly lower than average.[3]

The Octomom story touched a chord in the public's mind in part because of Kamrava's appalling disregard for the norms of his profession and in part because Suleman used the birth of these children to try to become a celebrity. But Kamrava was not the only one in his field to betray the norms of the profession. There was Cecil Jacobson in the 1980s, whose misdeeds, even though they were not tied to the new reproductive technologies, helped spur the Wyden Hearings. And when the Octomom gave birth, Ricardo Asch, on the lam in Mexico since the 1990s, remained a fugitive from justice.[4]

Kamrava, Suleman, Asch, and Jacobson, we want to be clear, were outliers. Willfully negligent physicians, women who use motherhood as a way to gain notoriety, eminent doctors who misuse their patients' eggs and embryos then flee the country, and respected practitioners who lie to their patients about their pregnancies for financial gain are hardly typical of those who are seeking to become parents and those who provide the services to make that possible.

Scandals are more than just sensational stories, however. Their broad coverage in the media tends to fire up the smoldering embers of American anxieties and ambivalence about the wide-ranging and far-reaching implications of these new ways of creating families. Over the years, the public—and our legislators and judges—have had arguments over the ethics of egg and embryo donation (or sale), surrogacy, the use of reproductive assistance by same-sex couples and unmarried men and women, racial and socioeconomic disparities in access to care, IVF for women past the age of menopause, elective egg freezing by women seeking to preserve their fertility in the event they fail to have children during their prime reproductive years, the creation of "three-parent" embryos, and more. Some issues have been resolved to one degree or another, other long-standing controversies still smolder, and new disagreements arise. One area of ongoing contention, which is far from resolved, is whether there should be mandated health insurance coverage for infertility services and assisted reproduction. Some say, "of course." Others maintain that the country ought to address the basic medical needs of all Americans before spending its health care dollars on the promotion of pregnancy. Years of debates in the media and disputes in the courthouse and legislative chambers have not—at least so far—led to any national consensus on how to resolve these issues.[5]

The 1975 moratorium on federal funding for embryo research, the recommendations of the Ethics Advisory Board in 1979, and the multiple commissions that followed all represent failed regulatory efforts. In 1978, when the Ethics Advisory Board was formed, researchers and clinicians in reproductive medicine appeared ready to accept federal guidelines because they would allow for the development of the new technologies through the mechanism of federally funded, peer-reviewed research. After all, this was the environment in which medical scientists had worked since the federal government became the major funder of biomedical research in the post–World War II era. But then the recommendations of the Ethics Advisory Board were ignored, the funding ban continued, and researchers and clinicians in this new field proceeded to develop the new technologies on their own. Soon, they grew reluctant to support governmental regulation of their research and practice.

In the ensuing decades, increased political polarization around the inter-related issues of abortion, gender roles, embryo research, and the expansion of the uses of reproductive technology made it more difficult for Congress to develop any sort of consensus. As a result, the United States, riven by the abortion wars, divided by arguments over the relationship between gender equality and reproduction, and split, in fundamental ways, over such basic questions as what it means to be a family and who should have access to the technologies of reproduction—has by default allowed the market to deter-mine the development of and access to assisted reproductive services.

We know that some people, including many physicians working in the field of assisted reproductive technology who find themselves being asked to pro-vide more—and more detailed—data for their annual reports on the success rates of their fertility centers, will take exception to our characterization of an unregulated reproductive marketplace by calling our attention to the *Fertility Clinic Success Rate and Certification Act of 1992* (also known as the Wyden Act). Most reproductive endocrinologists, including all the physicians to whom we spoke while writing this book, obey that law. Some of them had a hand in its creation. Because they follow the rules, they do in fact *feel* regulated, and every year, it seems, they are asked for ever more, and ever more detailed, informa-tion.[6] Nevertheless, it remains true that there are no legal penalties for IVF clinics or programs choosing not to provide that information. There is consid-erable peer pressure from the Society for Assisted Reproductive Technologies for compliance, but peer pressure is not the same thing as legal repercussions. The Wyden Act is consumer protection legislation that allows potential pa-tients to learn about a clinic's success rates in order to find one most suited to their particular needs.[7] It is a useful law, but it is not regulation.

The American Society for Reproductive Medicine, absent a regulatory system, has sought to manage the development and use of these technologies through voluntary mechanisms. The ASRM has a number of standing com-mittees, including one on practice and one on ethics. These committees take their duties seriously and regularly produce guidelines and opinions. The so-ciety has taken positions on nearly all matters affecting the field, including, for example, comprehensive insurance coverage for infertility treatment, including assisted reproductive technology, which it supports, and research into somatic reproduction, or reproductive cloning, which it opposes. Many of its members do pay attention to the ASRM's opinions and guidelines, and we discuss one of the organization's important successes—the significant reduction in multiple births—later in this chapter. Most fertility clinics also followed the ASRM's guidelines on elective egg freezing, which had initially designated the procedure as experimental. As a result, most clinics did not encourage the practice. Only after the lifting of the "experimental" designation did elective egg freezing become more common. The power of the ASRM's advice in this area had limits, however. Even as it removed the experimental label from elective egg freezing, the ASRM urged caution in offering the procedure, advice that was not heeded.

The leaders of the ASRM, including members who serve on committees and develop the organization's guidelines, are committed to ensuring that research and practice in the field are conducted in a scientifically and medically sound as well as ethical manner. But as conscientious as the ASRM seeks to be in its monitoring efforts, it is both an interest group and an organization with no real enforcement power beyond peer pressure. Membership in ASRM can be revoked, and a fertility center can be expelled from SART, but unless these centers are violating federal or state law, or physicians are engaging in actions that would result in their being stripped of their medical licenses, they can keep on treating patients. Even before he became notorious as the physician who implanted those twelve embryos into the Octomom, Michael Kamrava was ignoring ASRM guidelines, but how would anyone know? Professional guidelines and official opinions are clearly necessary, but they are not always sufficient.

Even though Congress has been famously unable to regulate assisted reproductive technologies, the federal government has had an impact on the development of the field in several ways nonetheless, by making possible the creation of a virtually unfettered marketplace for the provision and use of the new technologies. Congress has also used its power of the purse to prevent federal spending on research of which it does not approve through the mechanism of riders to appropriations bills. But a *funding* ban means exactly what it says. The research itself is not banned and still goes on. Nevertheless, as Christos Coutifaris, chief of the Division of Reproductive Endocrinology and Infertility at the Perelman School of Medicine at the University of Pennsylvania, told us in conversation on March 19, 2018, there are enduring consequences to having prevented the most important independent funder of biomedical research in the nation—the National Institutes of Health—from supporting grants in this area. One ongoing issue, he says, is the establishment of "a standard of practice without appropriate clinical trials."[8]

The original funding moratorium had come from the executive branch of government, but when President Bill Clinton, a Democrat, formally ended the ban in 1995 to allow for federal funding of stem cell research on existing embryos, the Republican Congress restored it by means of the Dickey-Wicker amendment to the annual appropriations legislation for the National Institutes of Health in 1996. The two men for whom the rider is named are now long gone from the House of Representatives, but the amendment lives on, inserted into the appropriations bill every year since. More recently, as we discussed in some detail in chapter 7, the US Food and Drug Administration, through its power to regulate gene therapies, was prepared to consider approving applications for clinical research on mitochondrial replacement techniques. (Because of Dickey-Wicker, the research itself would have been paid for by private funds.) At least one member of the Institute of Medicine committee anticipated that the FDA's consideration of such research would open the door to a larger role for the federal government in regulating assisted reproduction more broadly.[9] We will never know if he was right. In the end, through a rider to the appropriations bill funding the FDA, Congress prohibited the agency from using federal funds to review applications to employ this technique, thereby tying the agency's hands.[10]

A Proposed National Action Plan on Infertility

The most recent effort of the federal government to develop a national consensus on dealing with the problem of infertility and the use of assisted reproductive technologies began in 2010, with a declaration from the Centers for Disease Control and Prevention that infertility and its associated reproductive disorders constituted a serious national public health problem. The agency carefully specified that it was not seeking to make regulations or create federal policy; its goal was instead to foster an environment where policies could eventually be created. The CDC intended to create an action plan that would provide "federal and other government agencies, professional and consumer organizations, and other partners and stakeholders, a foundation and platform to work together to decrease the burden of infertility in the United States."[11] The agency released a draft plan in 2012, invited and received public comment, and unveiled its final version two years later. Systematic, regular data gathering in three areas formed the heart of the plan, which sought to measure the prevalence of the disease of infertility and its risk factors, assess existing infertility treatments and access to them, and examine the "economic and financial aspects of service delivery."[12]

All these things should have been measured all along, we might say, but starting now is better than never having this information. In its meticulously parsed phrases, the *Action Plan* made clear that the CDC was interested in more than information on medical conditions causing infertility. It also wanted data on the vast array of treatments, from medical to surgical to assisted reproductive technologies. This apparently innocuous recommendation—who could complain about gathering knowledge?—would, if implemented, provide considerably more information than is obtainable from the current SART/CDC database, which has statistics only on the assisted reproductive technologies. The CDC also wanted to collect data, for example, on "the safety and efficacy of the use of donors for infertility management (e.g., oocyte donation, oocyte cryopreservation, sperm donation, reproductive tissue donation, gestational surrogacy)" and to evaluate the effects of these treatments on "donors, recipients, and children conceived." This would be a major step forward in understanding the extent and impact of gamete donation. Because the use of an egg donor requires the use of IVF, we know exactly how many babies are born from donor eggs. But because artificial insemination is not considered assisted reproductive technology, neither the CDC nor any other governmental agency collects data on the numbers of children born as a result of sperm donation.

The *Action Plan* also called for research into the social and ethical aspects of reproductive medicine, including the status of efforts to eliminate racial, economic, and geographic disparities in access to care. The CDC did not call for new federal legislation to achieve any of these goals. Instead, it sought to build a "platform to stimulate discussion and collaboration among Federal agencies, professional organizations, academic institutions, and those who represent consumers of health services."[13]

If fully implemented, the recommendations detailed in the *Action Plan* could for the first time provide the nation with important information about children born through the use of three-party reproduction and shed light on the true extent of disparities in access to treatment. But thus far we have not seen much movement on the CDC's proposals. We suspect that even such a measured step may be too controversial and too contrary to entrenched and long-standing attitudes to succeed. The CDC developed these recommendations during the Democratic administration of President Barack Obama. The intent was to garner broad public support and have a real impact on public policy. Under the Trump administration and a polarized political climate, we have little faith that much, or perhaps any, of this action plan will go forward.

Have we come full circle? In 1979, the reasonable, considered advice of the Ethics Advisory Board, whose members came from different and in some instances opposing ethical and religious perspectives, could have been the first step toward regulation of in vitro fertilization and the other technologies that emerged from it. But their advice fell on deaf ears. Then, more than thirty years later, the CDC, defining infertility as a public health issue, sought to return assisted reproduction under the umbrella of the federal government. To be talking about "first steps" three decades apart, when every other developed nation has been monitoring and regulating these technologies for nearly that long, gives us pause as we seek to use the history we have just written not only to take a long view of the past but also to look toward the future. We are now four decades removed from the decision of the administration of Jimmy Carter to ignore rather than to grapple with the ethical and political ramifications of the creation of human embryos outside a woman's body. The technologies have multiplied. So have the medical and ethical questions about them.

Reducing the Incidence of Multiple Births after IVF: A Success Story

We begin with a problem that voluntary guidelines and professional leadership have addressed successfully. Multiple births are dangerous for both mothers and babies, but in the early years of IVF in the United States, no one seemed to consider it a problem. The first few births in England from the new technology had relied on a patient's natural cycles: a woman ovulated, produced one egg, and that egg was removed via laparoscopy, fertilized, and transferred into her uterus. One embryo, for the most part, meant one baby. Ovulation-inducing drugs soon became standard, and in the early years, doctors knew little about how many embryos they needed to transfer in order to have a successful pregnancy. As a result, they tended to transfer into the uterus all the embryos that, as several early practitioners in the field told us, "looked good." A few of these doctors remembered that their usual limit was four or five. Others said they told the patient how many they had and left the choice of how many to transfer up to her. There was always a sense of excitement in the press when twins were born: first twins born to the Joneses, first twins in Philadelphia, first twins in Washington, DC, first twins wherever. Mothers appeared on the news fresh from the delivery room, exhausted, with fathers proudly looking on, babies in their arms. Most of these early multiples were twins. In 1990, 28 percent of the deliveries after IVF were multiple births—24 percent of women delivered twins, and 4 percent delivered higher-order multiples, most of them triplets. (See figure A.3.) There was no organized effort to reduce their incidence.

Robert Edwards and Patrick Steptoe tried to sound the alarm on multiple births as early as 1984. Responding to reports of births of triplets and quadruplets in England, they argued that no more than two to three embryos—at most—should be transferred after IVF.[14] They were alone at the beginning, but others soon joined them. Within a few years, IVF programs in Britain and other European nations were prohibited from transferring more than three embryos. In the United States, even as late as the early 1990s, the situation was different. Wanda remembers that she and other obstetricians considered multiple fetuses to be a side effect of fertility treatments. A woman pregnant with triplets or more would be referred to a high-risk specialist, who would suggest that the couple consider fetal reduction, usually to twins. This was extraordinarily hard on the patients. Imagine having spent years trying to get pregnant and then being told that, for the sake of the survival of two of those fetuses, you should abort the other one or two of them.

It was a difficult issue, but for the most part, American reproductive specialists preferred not to address it. Howard Jones was virtually alone in the 1990s when he began urging his colleagues to limit the numbers of embryos transferred, although he was unwilling to specify a number, leaving it up to the experience of each physician to decide. Jones was soon joined by Eli Adashi, an expert in the problems of premature and multiple births. The two men called on the profession to act, and Adashi became the most prominent American reproductive endocrinologist to advocate for the reduction of the numbers of multiple births arising from IVF and ovulation induction treatments. Soon, more of his colleagues joined him.[15]

But even as practitioners were becoming increasingly aware of this problem, the public continued to think of multiple births as an interesting curiosity. "Is New Jersey Triplet Nation?" asked one *New York Times* reporter in 2005, as she counted three sets of triplets in one affluent southern New Jersey town and seven in another.[16] First-person accounts, however, told a different story. Suzanne Sanchez and her husband had spent "years and tens of thousands of dollars trying to conceive a baby," only to find out, once they succeeded, that she was carrying triplets. While most of her friends greeted the news with "joy," she said, congratulating her for getting "it all over with one shot," her obstetrician laid out the risks to the babies. Triplets, he said, were generally born prematurely and "with a lot of complications." The mother faced risks, too. Women with multiple gestation have higher risks of postpartum hemorrhage and a significantly higher risk of pregnancy-related death. The doctor suggested she consider fetal reduction, and Sanchez remembered thinking, "I'm a feminist who believes in abortion rights, [but] this was not the choice I had in mind."[17] In the end, she and her husband decided against the reduction. She endured multiple complications during her pregnancy and delivered at twenty-six weeks. Each of the babies weighed less than two pounds.

The birth of these babies and their subsequent care were covered by their own health insurance, plus Medicaid for the babies because they were under "2 pounds 10 ounces" at birth. The cost of their care reached $1 million. "Our quest to have a family," she wrote, "resulted in a significant financial drain on society's resources." Not every set of triplets has so many problems, but triplets are almost always born prematurely. Twins also have a higher risk of premature birth. Complications from multiple births can include problems with breathing and eating. Preterm babies have a greater risk of cerebral palsy, and other health problems can emerge later in life.[18] Sanchez's triplets were born in 2003, when multiple births were declining, although slowly, from their highs in the 1990s. In 1995, multiples accounted for 37 percent of deliveries. In 2005, it was 32 percent. Gradually, pressure from leaders in the fields such as Adashi, who made the decrease of multiple births one of his signature issues, began to make a difference. By 2015, just 19 percent of deliveries after IVF were multiples, and nearly all these multiple births were twins. There may have been a few higher-order multiples we could not count because of the way the data were reported, but this is nevertheless a significant decline (figure A.3).

The successful campaign to reduce the incidence of multiple births after IVF began with the determination of a few highly regarded physicians who persuaded the ASRM leadership to make this issue a high priority of the field. Practitioners heeded their advice, particularly when over time the data showed that transferring fewer embryos did not lower success rates, a persuasive argument in a highly competitive IVF marketplace. And finally, between the advice of their doctors and the data showing that success rates were not negatively affected, patients were choosing to have fewer embryos transferred. Among younger women, single-embryo transfer has taken hold, and there has also been a reduction in the numbers of embryos implanted in older women. Fertility centers now offer elective single-embryo transfer (eSET). In 2012, about 15 percent of women under thirty-five chose that option. By 2015, it was up to 35 percent.[19]

Professional guidelines have been effective in reducing the numbers of multiple births after IVF. Another problem remains, however. IVF is not the only cause of multiple births. The use of ovulation-induction medications— fertility drugs—about which no statistics on prescriptions are kept in the United States, is the leading cause of this problem. Fertility drugs can be ordered by any physician, and there is currently no way to know where and how they are being used. Adashi noted that including this information on birth certificates when a multiple birth occurred would be helpful but difficult to implement. Should the CDC's action plan ever be implemented, such data, relevant to multiple births from all fertility treatments, would be collected under its auspices—another reason to support the agency's efforts.

Three-Party Reproduction, Diagnosing Embryos, and Preventing Disease

Professional guidelines, followed by patient education, succeeded in reducing multiple births after IVF. Other issues are less amenable to such an approach. One of the most troublesome for many has been the overt commodification of eggs, sperm, and even embryos. We are aware that the term "commodification" has been applied by historians to children in the United States ever since they were no longer needed to till the field or round out the family income, but the ways in which sperm and eggs are marketed today, the creation of ready-made embryos, and concerning stories of intended parents rejecting offspring carried by gestational surrogates seem to us to take commodification to a different level.

Glitzy catalogues offer sperm that might produce every kind of potential baby. European Sperm Bank USA has "Abbott," a 6-foot, 1-inch mechanical engineer with blue eyes and light brown hair. Or you could have "Arroyo," a 5-foot, 6-inch Hispanic university instructor with brown eyes and black hair, or "Quintin," a 5-foot, 8-inch African American / Lithuanian business student. Its website contains a "products and prices" page.[20] Some sperm banks have end-of-year sales. With egg freezing proving to be successful, egg banks on the sperm bank model have become more common. The availability of frozen eggs and frozen sperm also makes possible the creation of ready-made embryos, which are now available from one enterprising physician, Ernest Zeringue, whose clinic we considered in chapter 7.

Compensated egg and sperm donation have long been considered by most practitioners to be reasonable and ethical reproductive options. The idea of creating embryos essentially on consignment, however, takes commodification to a more ethically suspect level. In each of these cases, as sociologist Rene Almeling reminds us, "It is *family* that is for *sale*." In egg donation, however, the financial transaction is softened by framing the exchange as a "gift, a compensated gift, but a gift" nonetheless. Such framing makes it possible "to manage the cultural tension of women being paid for eggs that become children and create families." Almeling's work uncovers the gendered market for human gametes, where sperm can be seen as a product for which men are paid, but eggs are a priceless, if remunerated, gift.[21] To what extent the new frozen egg market will change this rhetoric is still to be determined, but once a woman is no longer offering a compensated gift to a particular couple but selling her eggs to Cryos USA, which will put those eggs on their website to be added to someone's shopping basket, it seems inevitable that the process will increasingly resemble sperm donation.

Where does the sale of embryos fit into the existing marketplaces for eggs and sperm? Many in the field find the idea distasteful. Perhaps that is why, so far at least, Zeringue's clinic appears to be the only one openly engaged in such a practice. But we wonder how different—in moral terms—ready-made embryos are from those being offered by evangelical Christians who use the inconsonant language of "embryo adoption" to refer to a process where money changes hands so that like-minded infertile Christian couples can acquire embryos that have been given to the agency. They do not seem to recognize that they too are engaged in a commercial transaction. It remains to be seen whether the ethical issues arising from the selling of embryos can be addressed solely with ethical guidelines. We do know that guidelines alone cannot deal with the callous rejection of children created by surrogacy in the event that the commissioning parent(s) find the baby defective, as we saw in chapter 7. Does the fact that a person or couple is purchasing eggs, sperm, or embryos, or paying for a gestational surrogate, cause them to believe they have a right to a certain outcome? And if they don't get that outcome, is it acceptable for them to reject the child in the same way that they return an unsatisfactory purchase? Sometimes they believe that they can, and they do.

Sperm donation has a long history of enabling women with infertile husbands to bear children. In addition, in at least some doctors' offices, unmarried women (who may or may not have been partnered with another woman) were able to access donor sperm in some parts of the country as long ago as the 1930s. By the 1980s, providing the service to lesbian couples became more common.[22] With the development of intracytoplasmic sperm injection (ICSI), which allows men with severe infertility to father a child with their own sperm, the use of donor sperm among heterosexual couples declined. Nowadays, lesbian couples and unmarried women have become the principal markets for this product. Paid egg donation became a more common phenomenon in the 1990s and continued to grow in the first decade of the twenty-first century. And although there is controversy over what is considered appropriate compensation for egg donation, it too is an accepted part of modern reproductive medicine.

Gestational surrogacy became available in the mid-1980s, when Wulf Utian's IVF clinic offered what he called a "host uterus" program. Surrogacy remains uncommon, but even though it takes up a small sliver of the business of assisted reproductive technology, it is a profitable sliver in many metropolitan markets. As we were working on this chapter one morning, one of Margaret's graduate students came into her office to tell us what happened at a Philadelphia GayBINGO event he attended over the weekend. GayBINGO is a monthly fundraising event to support the Philadelphia AIDS Fund. Each individual game is sponsored by a business—bars, coffee shops, cultural venues, for the most part—and between games, the sponsor offers a pitch for its organization. As it turned out, selling its services along with Starbucks and the Sawtown Tavern was Reproductive Medicine Associates and its local partner, Jefferson Hospital. They sent a representative to the games to let Gay-BINGO attendees know about their targeted services, including surrogacy, for same-sex couples.

Same-sex male couples and women who do not have a uterus or are otherwise unable to undergo a pregnancy are the primary markets for gestational surrogates. The stories of single men seeking surrogates so that they can have children are extremely rare in the United States, and generally emerge only when the situation ends badly. There was the Easton, Pennsylvania, man who fathered a son through surrogacy in 1995, killing the baby five weeks later because the infant was crying. More recently, in 2017, a surrogate carried and gave birth to triplets for a fifty-one-year-old California man who lived in his parents' basement, to which he brought the babies. According to the man's family, he already had a history of animal abuse and was now neglecting the children. Distraught, the surrogate sued for custody on the grounds of "abandonment, forcing the babies to eat off a basement floor and diapers so filthy that the three boys had to be taken to a hospital." The man's sisters confirmed the last two of the allegations, but the surrogate lost the case because, according to California law, he was the legal father, and she had no standing.[23] Such awful stories make headlines because they are rare, just as were the ones we told in previous chapters. When surrogacy goes well, there is no story to tell.

There are other troubling aspects to surrogacy, however, including what seems to us to be an insufficient understanding of the fact that pregnancy can be dangerous. We have also been concerned to read about the overtly transactional behavior of some contracting parents. Although it was touching that Alex Kuczynski's surrogate, Cathy Hilling, remembered Alex's birthday with a present and brought *her* a gift on the day *Cathy* was the one in labor, we were struck by the fact that Alex never thought to reciprocate. Equally poignant was a story told by another surrogate, who had a baby for a male couple who treated her with kindness and generosity during her pregnancy and delivery. She was touched by how much they seemed to care about her, and when they asked her to serve as their surrogate a second time, she agreed. When her first cycle had to be canceled because of what seemed to be a temporary medical issue, the doctor prepared to try again the following month. But the potential fathers did not want to wait and immediately found another surrogate. The surrogate thought she had a relationship with them. They apparently did not feel the same way. "They didn't mean to hurt me," she says. But hurt her they did.[24]

The United States is virtually the only developed nation that allows for compensated surrogacy. Australia, Canada, Denmark, Great Britain, New Zealand, and Sweden allow altruistic surrogacy. Some countries—such as

France, Germany, Italy, Norway, Portugal, and Spain—prohibit surrogacy entirely.[25] India and Thailand, two countries that had been popular destinations for couples seeking surrogacy for more than a decade, have sought to end international surrogacy in their countries. Thailand suffered several scandals, and India had earned a reputation as a surrogacy factory. In 2016, India's foreign minister, Sushma Swaraj, told CNN that "there were too many instances of people abusing surrogacy, including couples who had abandoned unwanted children or babies born with disabilities." Thailand's laws have already been changed. India has a bill under consideration banning international surrogacy, and meanwhile it has tightened visa restrictions so that foreigners cannot come into the country for the purpose of surrogacy. While those seeking low-cost surrogacy are now heading to the nations of Georgia and the Ukraine, well-to-do Europeans and Asians are traveling to the United States. According to the most recent statistics we have for the percentage of foreign patients using gestational carriers in US fertility centers, in 2013, that figure was 18 percent. It is likely higher today.[26]

In contrast to the American approach toward surrogacy, the United Kingdom has comprehensive laws that seek to strike a balance among the rights of all parties to the agreement. There are about three hundred births to surrogates every year, and compensated surrogacy is prohibited; only altruistic surrogacy is legal. Surrogates are, however, entitled to reimbursement for medical and related expenses that can range from £7,000 to £15,000 ($9,200 to $19,700). The surrogate, as the birth mother, is the legal mother of any children born through surrogacy until there is a signed order from the courts transferring legal parenthood to the intended parents. This is true even if the surrogate has no genetic relationship to the child. In 2018, the Law Commission began a review, expected to take three years, of the current laws, with a special focus on the length of time it takes to transfer parenthood from the gestational mother to the intended parents.[27]

No law is perfect, but while researching and writing this book, we have come to believe that surrogacy for hire creates an inherently unequal and potentially exploitative relationship. If we extend Rene Almeling's argument about compensated egg donors to compensated surrogates, it seems clear that many women who agree to be paid to bear a child for someone else have internalized a belief that they are offering a gift to the intended parents. But it is nevertheless a commercial transaction—my body for your baby. And my body may suffer complications from pregnancy. I may miscarry. I may be carrying triplets when you want just one baby. We may find out late in pregnancy that there is something wrong with the baby, and you may simply decide to walk away. Or I may discover that you are an abusive father or a negligent mother.

We also know that without surrogacy, many loving couples would not have children. The children of Doug and Eric, about whom we wrote in chapter 7, likely couldn't have asked for better parents, and their surrogate felt valued for her contribution to their family. Or consider Susan, also from chapter 7, whose medical condition would never have allowed her to undergo a pregnancy. We have a great deal of sympathy for couples in these and similar situations. Great Britain's approach to surrogacy seems to provide for a more equitable process; altruistic surrogacy based on its model could help equalize the relationship, making it less overtly transactional and limiting potential harm to both the surrogate and the children. It is also possible that potential technological developments in reproductive medicine itself—uterus transplants and external artificial uteruses—to which we turn later in the chapter, may provide alternatives to surrogacy. Both, however, have other drawbacks.

Diagnosing, screening, and altering human embryos have also caused controversy in the twenty-first century. Techniques such as preimplantation genetic diagnosis (PGD), preimplantation genetic screening (PGS), and mitochondrial replacement techniques (MRT) make it possible to diagnose hundreds of genetic diseases in an embryo, screen embryos for chromosome abnormalities, and prevent mitochondrial diseases from being passed on from mother to baby. Some believe that the use of these technologies will lead to embryo selection for traits considered socially desirable—this is often referred to as the creation of designer babies—and therefore oppose them.[28] Disability rights advocates have argued that the desire to have a child free of a genetic disease could lead to the devaluation of the lives of those who are living with a disability. This concern should be taken seriously. As a society, we must ensure social policies in which people living with a disability now, and everyone who may suffer a disability at any point in their lives, are valued. But we also believe that it is a doctor's role to prevent suffering. To use PGD to select an embryo free of cystic fibrosis or sickle cell anemia, or to use MRT to prevent the transmission of a devastating mitochondrial disease, we would argue, is to prevent suffering, not to create a designer baby.

Today, scientists in China, Great Britain, and the United States are studying the use of the gene-editing technique CRISPR/Cas 9 to correct a defective inherited gene in human embryos. After two years of study, an international commission created by the National Academy of Sciences and the National Academy of Medicine recommended that countries choosing to conduct such research should follow strict guidelines. "Editing the DNA of a human embryo to prevent a disease in a baby could be ethically allowable one day," reported *Science Magazine*, "but only in rare circumstances and with safeguards in place."[29] Currently in the United States, however, such work is illegal under the same appropriations amendment that prohibits funding for MRT. As a society, we should be addressing the development and use of these technologies through national commissions and subsequent legislative processes. Instead, the only laws we have that deal with embryo research are riders to annual appropriations bills.

Motherhood Postponed

"A woman in her late fifties was just referred to me," one of Wanda's colleagues, a maternal fetal medicine specialist who cares for women with high-risk pregnancies, told us out of the blue one evening. We had asked to talk to him about complications women face when pregnant with twins or higher-order multiples, but our talk soon turned to broader issues. This woman wanted to have a baby, he said, and the fertility center where she sought treatment asked for his judgment on whether it would be safe for her to carry a pregnancy. While we were asking Robert Debbs about the health risks of carrying twins and triplets, he had apparently been thinking about another aspect of our research for this book. "How old is too old to have a baby?" Debbs was being asked. The fertility clinic was trying to decide on what side of the divide to place this patient. "She seems very fit," Debbs continued, "She looks like she's in her late thirties." He told us that he intended to make sure she was thoroughly evaluated for a range of health issues before giving the clinic his advice.[30]

As we have shown in this book, women are having their first children at an older age than women of earlier generations. They may be in their thirties or even forties when they enter the delivery room, but most women are not choosing to forego motherhood. Just 15 percent of women aged forty to forty-four have never had children. Nevertheless, if a woman seeks to become pregnant using IVF in her forties, it is difficult to conceive with her own eggs. At forty-five and older, her chances are only around 1 percent.[31] However, since the discovery in the 1990s that women in their forties and fifties can successfully conceive with donor eggs, increasing numbers of such women have been taking that option. In the 1990s, Mark Sauer, Richard Paulson, and Rogerio Lobo, who were among the earliest to make that discovery, were enthusiastic about treating women up to the age of fifty-five, describing these women's pregnancies and births as relatively uneventful.

Two decades later, Mark Sauer changed his mind after reevaluating those earlier studies. Even though the first cohort of seven women in their forties whom he treated underwent rigorous screening before being admitted to the program, of the six who conceived, every one experienced difficult pregnancies or births. Only one had a vaginal delivery, and that was a fetal demise. Another had twins. The others had a range of complications. In another study conducted by Sauer, this one of 101 women who underwent IVF between the ages of fifty and fifty-nine, the complications were even more serious. One of these women died in the first trimester, others had hypertension or diabetes, and 87 percent had cesarean deliveries. While Sauer was not willing to say that such women should be denied fertility treatment, he wanted them to be well informed about the risks.[32] In 2015, there were 754 births in the United States to women over age fifty. That number is likely to rise. The earlier cutoff age of fifty-five for acceptance into an IVF program is now being breached. Reviewing these data persuades us that women in their fifties should not simply be informed about the risks—they should be actively discouraged from seeking treatment. There comes a time when a woman is too old to have a baby, and we believe that the reproductive medicine community should be willing to say so.

The circumstances are different for women in their forties, even though they, too, have more complications than younger women. Given the general upward trend in first births, we believe that the numbers of women in this age group who seek treatment at IVF centers will continue to rise. We also know that women in their forties have always had babies, although in centuries past those babies were usually their last, not their first. Such women, we believe, should be fully advised of both the general risk factors for their age group and any specific concerns relating to their own health so that they can make an informed decision. It is simply not possible, in the United States today, for many women to consider motherhood during their most fertile years. Significant social, cultural, and political changes—equal parenting by husbands and partners; flexible work arrangements that are not viewed as mommy-track jobs; and universal, affordable, and excellent child care—would be required for more American women to be able to have both careers and a family, or even to marry, at a younger age. There are currently no signs that the nation is moving in that direction. Women who wish to have both a fulfilling professional life that may require long years of training or education and children face a genuine dilemma. Men in this society do not have to make such choices, and women should not have to, either. But they do, and the increased use of fertility services reflects their dilemma.

It is important for women to have correct information about fertility and age, and gynecologists can provide it. It is not helpful, however, to tell them that they ought to have their children in their twenties, which was the conclusion Mark Sauer arrived at after his retrospective assessment of the risks of later childbearing among the women in their forties whom he helped to become pregnant. Society, Sauer argued, needed "to promote earlier efforts at procreation, while countering myths suggesting you can 'have it all' by delaying reproduction until a time that is convenient." He realized, he said, that "a social reengineering back to a conventional time may be difficult, if not impossible to do, but a failure to do so will result in increasing numbers of women left childless and without adequate medical interventions to reconcile their needs."[33] This attitude puts the onus on women. We see not a word about reengineering marriage as a more egalitarian partnership, or reengineering the workplace and the larger society to provide the kinds of structural supports that would allow women to have their children earlier and still have a thriving career.

Realistically speaking, a diminution in the numbers of women who choose to build their careers before having children, whether they marry or not, is unlikely. Until recently, if a woman failed to become pregnant with her own eggs, her only choices were childlessness, adoption, or donor eggs. Now there is also elective egg freezing. The distinguished Yale anthropologist Marcia Inhorn caused a stir a few years ago when she wrote that she wished she could have had that option when she was trying both to succeed in her profession and have children. She was not extolling freezing one's eggs as fertility insurance or a substitute for needed social changes. "Promoting egg freezing as a quick-fix technological solution," she insisted, "does not solve the unfavorable employment policies that cause women to lean out of their careers." Nevertheless, she said, she had decided to encourage her graduate students to consider doing so as an option for the future. Having struggled herself with the difficulties of balancing a high-powered career with a desire for children, Inhorn was trying to be practical. She had a stillbirth and a miscarriage in her mid-thirties before having two children, one at age thirty-seven and her second at thirty-nine. She cared about her graduate students and knew what they could be facing a few years hence.[34]

Her advice, however, engendered considerable controversy. Challenging Inhorn, a group of young scholars argued that egg freezing frames the difficulty of balancing career and family "as an individual problem," implying that "there is little perceived need and less support for structural changes."[35] Feminist scholars Lynne Morgan and Janelle Taylor also argued against the idea. "The truth behind feminist struggles is that achieving a more just society will require paid parental and sick leave, affordable child care, comprehensive health insurance, immigrant health care, and adequate wages." A frozen egg, even if it works, does not solve the problem of gender inequality.[36] But what does a woman do about her own desires as she seeks to achieve a just society? Here we are, a century after women won the right to vote and more than half a century after the passage of the Equal Pay Act, and women have not yet achieved equality.

The disparities are evident in the workplace, where women earn, on average, about 79 cents for every dollar earned by men, and the injustices don't stop there. Black and Hispanic women make even less than that, and women of all races face a "mommy penalty." Mothers make 3 percent less, on average, than women without children. Perhaps unsurprisingly, the opposite is true for men. Fathers make on average 15 percent more than men without children. As one report interprets the data, fathers are viewed as more responsible employees than those who don't have children. In contrast, mothers are viewed as more committed to their family than to the workplace and therefore as less responsible employees than those without children at home.[37] Inhorn's message seemed to be that if egg freezing can give some of these young women the opportunity to postpone childbearing, they should consider it. But those who criticized this advice are also correct. Individual solutions to a problem are not a substitution for social justice. And besides, when it comes to assisted reproductive technologies, even individual solutions are out of reach for many who live in this country.

Writer Reniqua Allen was unmarried, childless, in her thirties, and thinking about her future. Should she freeze her eggs? Attending an information session in New York, she was surprised to see that just she and one other woman were black. Maybe it was like this everywhere, she thought. Allen's instincts were on target, said Desiree McCarthy-Keith, a prominent African American reproductive endocrinologist who practices in Atlanta. "Black women are left out of the conversation around egg freezing," McCarthy-Keith told her. "Historically, fertility treatments have been mostly targeted to and used by white women, middle-class women, so in the initial presentation of fertility treatments, they didn't really include us in the conversation." In the end, Allen said, she still might freeze her eggs, but she also would like the "fertility industry" to "realize the narrowness of the lens that it's using to talk about this technology." Black women, too, she said, should be able to "feel like egg freezing isn't just for their rich white peers and to know that we, too, can make unconventional decisions the norm."[38] That's if they can afford it. One round of egg freezing costs close to $10,000, and then those eggs will have to be stored, costing another $500 a year.

Allen felt alone and a bit unmoored as a black woman trying to decide in 2016 whether assisted reproductive technology should play a role in her life. She might have felt less so if she had known that First Lady Michelle Obama, when she was in her thirties, had wrestled with similar, if not exactly the same, issues. The Obamas married in 1992, when she was twenty-eight and he was thirty-one. She loved children and always wanted to be a mother, and once they felt more or less settled as a couple, they began trying to conceive. At first, they felt only disappointment, not worry, when she failed to become pregnant. Because Barack's duties as a state senator took him away from home four days a week, they thought, perhaps he was just not at home at the right time. After all, a woman can conceive only when she is ovulating. And then she did become pregnant, but their joy was short-lived. A miscarriage left them disheartened. Michelle told a few friends, who shared their own heartaches. One of them also recommended that Michelle see the infertility specialist who had treated her.[39]

The Obamas, it turned out, had unexplained infertility, a formal diagnostic category for a condition in which women or couples are unable to conceive for reasons that are not clear. Michelle was prescribed the oral ovulation induction drug Clomid, a routine first-line medical therapy in such cases. When this regimen failed to lead to a pregnancy, the doctor recommended IVF. She and Barack were "inordinately lucky that my university health insurance would cover most of the bill," she wrote. Actually, they were inordinately lucky that they lived in Illinois, which in 1991 had become just the fifth state in the nation to mandate comprehensive coverage for infertility treatment, including IVF. Without the law, her university might have covered her treatment, but it would not have been mandated to do so.

Although infertility affected them both, Michelle remembered, the burden of the treatment fell on her, and that felt unfair. She was the one injecting herself with fertility drugs, keeping multiple medical appointments, and rearranging her work schedule so that she could go to the clinic for daily ultrasounds to monitor her eggs. "He was doting and invested, my husband, doing what he could do," she wrote. Still, "his only actual duty was to show up at the doctor's office and provide some sperm. And then, if he chose, he could go off and have a martini afterward." She knew that "none of this was his fault, but it wasn't equal, either." As she thought about the difference, she found herself, she said, "in a small moment of reckoning. Did I want it? Yes, I wanted it so much. And with this, I hoisted the needle and sank it into my flesh."[40] She conceived and experienced an uneventful pregnancy. Their daughter Malia arrived on July 4, 1998. Their second daughter, Sasha, was also conceived using IVF, and she was born three years later.[41]

As First Lady, Michelle Obama had been reticent about matters involving her family. She never spoke publicly during her years in the White House about having used IVF, revealing her experience just before the publication of her memoir in 2018. She surely knew that black women are vastly underrepresented both in the media's representation of the problem of infertility and in the offices of those who treat it. Was she surprised, we wonder, that her story resonated so strongly in the media?

Michelle Obama's candor about having used IVF to become pregnant recalls the decision of an earlier First Lady, Betty Ford, who in 1974 announced publicly that she had been diagnosed with breast cancer during a screening mammogram, bringing out into the open a diagnosis that was so deadly for many women that they did not disclose the condition even to their friends or children. Ford went on to chronicle her surgery and recovery in interviews, and her example encouraged thousands of women to make appointments for breast examinations and screenings. And at a time when television was a far more reticent media, the *Today* show's Barbara Walters demonstrated for her viewers how to do a breast self-examination. She was clothed, of course, but it was still seen as a daring thing to do.[42] Will Michelle Obama's story resonate as strongly? It is too soon to tell, but by writing about her own experience with assisted reproduction, our first black First Lady may chart a new course for other black women experiencing infertility—and perhaps for all American women—encouraging them by her example to seek care.

Black women have been consistently underrepresented in fertility centers across the country, whether we are talking about egg freezing, IVF, other assisted reproductive technologies, or medical treatment for infertility, compared to affluent white couples and individuals, as are Hispanic women and white women with low and moderate incomes. We often wonder whether African Americans would have been more likely to see themselves as potential IVF patients had James Daniell's black patients, the Pattons, been the first Americans to have a baby after IVF instead of the Carrs. But of course we have no way of knowing.

At the end of the twentieth century, about 85 percent of all egg retrieval cycles were initiated in whites. Black patients accounted for just under 5 percent, and the others were Hispanic, Asian, or of other races. Little appears to have changed since then, at least among patients at centers in urban areas. According to a 2015 study, the Fertility Center of Illinois in metropolitan Chicago had a patient population that was about 74 percent white, 5 percent black, 13 percent Asian, and 7 percent Hispanic. These figures are similar to those in a study conducted almost a decade earlier at the Center for Reproductive Medicine at Brigham and Women's Hospital in Boston. There, about 81 percent of the patients were white, 5 percent African American, 4 percent Hispanic, and 10 percent Asian.[43]

Affluent couples, particularly affluent whites, are overrepresented in America's IVF clinics for several reasons. One is the cost. The median price of just one cycle of conventional IVF (in which a heterosexual couple uses their own eggs and sperm) in the United States today is about $19,000. That includes medications but not additional services such as ICSI or PGS. That's just one attempt, and many women do not conceive on their first try. The median household income in the United States today is just under $57,000. If a couple earns at or near that amount, that one cycle will cost about a third of their income. IVF is about three to four times more expensive in the United States than in Europe. And unlike European countries, in most of the

United States even today, there is little to no insurance coverage for the procedures. According to the ASRM, the use of assisted reproductive technology increases threefold in states that mandate comprehensive coverage for them, but only seven—Connecticut, Delaware, Illinois, Maryland, Massachusetts, New Jersey, and Rhode Island—have such coverage.[44] (See table A.1.)

Insurance coverage can increase access to care by making it possible for those with lower incomes to receive treatment without undue, and sometimes unbearable, financial hardship. It can also lower the incidence of multiple births by putting less pressure on patients to have more embryos transferred to increase their chances of a favorable outcome in one cycle. The ASRM and RESOLVE both strongly favor insurance coverage for IVF and all infertility services. ASRM also argues that physicians benefit as well. When patients have insurance, doctors can "provide care based on medically-indicated factors rather than what the patient can afford." Perhaps equally important, the ASRM Ethics Committee argues, insurance serves a social justice goal by "sparing physicians from having to turn away patients because of inability to pay."[45]

We are not persuaded that the profession needs to wait patiently for insurance coverage to advance social justice goals. There are a few intermediate steps that the organized reproductive medicine community could take. Some reproductive specialists volunteer for the International Council on Infertility Information Dissemination (INCIID), which provides donated IVF treatment to couples who meet a set of financial and medical criteria.[46] But the profession has the ability to do more. Profitable private centers, for example, could create charitable arms of their practices to offer care on a sliding scale. University-based centers might emulate the Center for Reproductive Health in the Department of Obstetrics and Gynecology at the University of California, San Francisco (UCSF), which established an IVF program at a county hospital with which UCSF is affiliated. Its program at San Francisco General Hospital is staffed by UCSF residents in obstetrics and gynecology under the direction of a senior fellow in reproductive endocrinology and infertility. The people served by this program are not only poor but also often marginalized by their limited command of English, low "health literacy," and undocumented immigrant status. This successful, although necessarily somewhat limited, program may work in other cities as well.[47]

Although the ASRM strongly endorses mandated insurance coverage, support is not universal among its membership, and the current political climate may have affected attitudes. Nearly all the physicians we interviewed or spoke with informally before the 2016 presidential election supported mandated insurance coverage for IVF. After the election, with Republican Donald Trump in the White House and his party holding majorities in both House and Senate, the overturn of the Affordable Care Act became a real possibility. Although the law itself was not repealed, Congress and the Trump administration have chipped away at its benefits. Perhaps unsurprisingly, the physicians we talked to in 2017 did not express the same level of support for coverage of IVF as an essential element of health care as did those we interviewed before the election. If we can characterize the doubters, the general refrain was this: We must protect a base level of health care for those who cannot access it now and may not be able to do so in the future. How can we justify pushing for coverage for IVF when so many Americans are in danger of losing access to all types of health care? If people want IVF, maybe they just should have to pay for it. There are more important things to worry about now.

We would make a different argument. Even understanding the threat to Americans' health care in general, the two of us believe that just as health care is a human right, fertility care should be considered part of a woman's basic health care. Although the Affordable Care Act was imperfect in many ways, the expectation of many of its supporters was that it would be expanded over time. Instead, at least in the short run, the reverse happened. Nevertheless, we remain convinced that fertility care, including basic IVF, should be a part of health care coverage. Studies have shown that in states where coverage is mandated, it adds only modestly to overall insurance costs. To make sure it is affordable, this coverage should not be open ended, could be age restricted (perhaps up to the age of forty-five), and could limit the number of cycles to the average number of cycles typically needed to conceive.[48]

We know that providing health insurance will not be enough to end all disparities in access to treatment. There are cultural, religious, and language barriers to consider. People who have had previous negative interactions with the medical community may be reluctant to engage it again, and doctors themselves may erect barriers. As the ASRM Ethics Committee insightfully observes, "Physicians may consciously or unconsciously make assumptions or possess biases about who deserves to be a patient and who wants or deserves treatment." The authors who reported on the outcomes of the low-cost IVF program at San Francisco General Hospital put the issue even more bluntly: "Tragically, the need for treatment of infertility in low-income populations . . . often provokes reactions of disbelief and discomfort among health care providers owing to a lack of empathy or misguided rationales such as overpopulation concerns."[49] Indeed. But surely one of the first steps we should take is to ensure that when people walk into the fertility center, they at least have insurance coverage.

Uterus Transplants and "Womb-Like Devices"

In late November of 2017, a woman gave birth to a baby boy at Baylor University Hospital. This might seem like a no-news story, except that this woman had been born without a uterus. If not for an unusual clinical trial underway at the university, it would have been impossible for her to bear a child. A local nurse, who heard about this experimental procedure on the radio, had altruistically donated her uterus to the woman, who became the first in the United States to give birth after uterus transplantation. Baylor is one of three medical centers in the United States holding clinical trials of this surgery, and as of this writing the only one that has enjoyed success. The Cleveland Clinic began trials in 2015 and the University of Pennsylvania in 2017. Christos Coutifaris, who had trained under reproductive medicine pioneer Luigi Mastroianni and at the time was the president of ASRM, told us that the program was important for a number of reasons, among them allowing women unable for cultural or religious reasons to use a gestational carrier to bear children.[50] Until the birth at Baylor, the only country in which the technology had succeeded was Sweden. By 2017, eight babies had been born there; all of their mothers, like the woman at Baylor, had used uteruses from living donors.

For heart and full lung transplants, it is not possible to use a living donor, but kidneys and livers can be transplanted from both deceased and live donors.[51] Doctors believed that this was also true for uteruses, but thus far the only successful births had come from using the organs of living donors. Then, in December of 2017, just a few weeks after the birth at Texas, a thirty-two-year-old woman in São Paulo, Brazil, gave birth from the uterus of a woman who had died suddenly after a ruptured brain aneurism. The donor was just forty-five and the mother of three children. Her family had already arranged to donate her heart, liver, and kidneys when they were asked if they would be willing to donate her uterus as well. They agreed. The recipient had previously undergone in vitro fertilization using her eggs and her husband's sperm. The resulting embryos had been frozen, and one of them was transferred into the transplanted uterus seven months after the surgery. The woman conceived, experienced an uneventful pregnancy, and was delivered by cesarean section at thirty-five weeks. The transplanted uterus was removed at the same time, which allowed the patient to discontinue the use of anti-rejection medications, drugs which those who have donated kidneys, hearts, and lungs must take for the rest of their lives. The team of physicians and scientists who made this birth possible did not publish their results until December of 2018, almost a year after the baby girl's birth, at which time they reported that the mother was in good health, and so was the baby, who was thriving and developing normally.[52]

Approximately fifty thousand women have uterine factor infertility. Some, including both of the women who successfully gave birth in Texas and São Paulo, were born without a uterus; others have one, but it does not function; and the remainder have had hysterectomies because of disease. These women have ovaries and produce eggs, which allows them to conceive their own children using in vitro fertilization, but they cannot carry a pregnancy. A uterus transplant would make it possible for them to do so, although it is still too early to determine how reliable uterus transplants will be. We do know they are likely to be expensive, possibly costing as much as half a million dollars. For now, the *New York Times* reports, because the transplants are being done as part of a clinical trial, for the women enrolled, "much of the cost [is] covered by research funds." Whether they would be covered by medical insurance if they were no longer experimental is unclear.[53]

The patients in Sweden had their uteruses donated by mothers, sisters, and in one case, a close friend. In Texas, the donor was an altruistic stranger. In Brazil, the decision to donate was made by the deceased woman's family. Kate O'Neill, who is the co-lead investigator for the University of Pennsylvania's uterus transplant program, says that it isn't clear whether transplants from living or deceased donors will be more successful. For that reason, Penn's program plans to do both. With a living donor, O'Neill told a reporter, the doctors have the time to make sure in advance that the donor organ is suitable, whereas with a deceased donor, the organ must be removed soon after brain death and quickly transplanted. One advantage of cadavers, however, is that "surgeons can take more tissue from the vagina and blood vessel network that is possible with a living donor."[54] Donors are not likely to be easy to come by either way, suggesting that even with continued success, this procedure is likely to remain limited.

Uterus transplants seem familiar in a way because most of the public is aware of kidney, heart, and lung transplants. In these instances, however, only the patient is exposed to medications to prevent organ rejection. With a uterus transplant, the fetus of the pregnant woman will also be exposed to some of these medications, which is of some concern. Are there alternative technologies to be considered? One possible substitute for an absent uterus, envisioned as early as 1932 in *Brave New World*, is an artificial, external one. The science has not come close to reaching a point where it is a serious possibility; it may be both undesirable and impossible to replicate the human uterine environment. However, researchers at the Children's Hospital of Philadelphia (CHOP) announced recently that they had created "an extrauterine system to physiologically support the extreme premature lamb," which was the title of their article in *Nature* reporting on their success. The researchers seem determined not to call it an artificial uterus. Instead, according to the headline for CHOP's announcement, it is a "Unique Womb-Like Device" developed in hopes of providing better outcomes for extremely premature babies.[55] The researchers have also expressed considerable doubt that this device or another like it could ever become an actual artificial uterus in which embryos could be implanted and gestated.

We trust them on their research. But we also recall a speech made by Howard Jones on the occasion of his one hundredth birthday, in which he urged his colleagues to take up research on what he called "pregnancy in artificial wombs." Jones was troubled by the use of what he called "the borrowed uterus" because of its potential for exploitation, and he was especially critical of the employment of women as surrogates in low-income parts of the world. He called it a form of "slavery," and he wanted to end it.[56] An artificial, external uterus, he said, would make it unnecessary.

Has the Regulation Train Left the Station?

We are completing this book in 2018, forty years after the birth of Louise Brown, the world's first IVF baby. In those forty years, nearly every developed nation has managed to implement national policies and regulations on the use of assisted reproductive technologies—but not the United States. In this country, ethics commissions come and go. Recommendations for national policies from bioethicists and others fall on deaf ears in Congress. A decade ago, business scholar Debora Spar urged policymakers to accept that reproductive medicine has become a business and agree to regulate it as such. "What the market needs" Spar wrote, "is a *politically determined* strategy, one that emerges from a dedicated and explicit political debate."[57] Whether we are seeking regulation of a business, as she would have it, or a solution to a pressing public policy issue, as we contend, it seems clear that the political arena must be our starting point.

When we embarked on this project six or so years ago, we thought we would be completing it in an era in which the right to health care would be expanding, not contracting, with the Affordable Care Act serving a first step toward universal health care. We anticipated debates over the kinds of treatments that would be covered and expected public dialogues about coverage to expand to include larger issues of oversight. We hoped that those discussions would include guidelines on the use of IVF and other assisted reproductive technologies. Sadly, we were wrong about the nation's trajectory toward universal health care, at least in the short run. But the short run is not forever, and once the trajectory changes, we urge the CDC to move quickly to implement its action plan on infertility. At the very least, carrying out that plan would provide reliable information on egg and sperm donation and gestational surrogacy, including data on the impact of these practices on everyone involved, from donors and recipients to the children conceived. Practitioners, potential patients, and policymakers would also have reliable information on egg freezing and its implications, disparities in treatment, and the social and ethical dimensions of new technologies.

Implementing the action plan could lead to a discussion of whether it would be possible for the United States to create a distinctively American version of Great Britain's Human Fertilisation and Embryology Authority. HFEA is an independent government authority that regulates both research and clinical treatment.[58] And although HFEA alone was not able to authorize lifting that nation's ban on the use of mitochondrial replacement techniques— that took an act of Parliament—the agency is now overseeing the responsible and limited use of MRT. If Americans could come to agreement on the development and use of these technologies, a wider policy discussion could follow.

Recently, Eli Adashi and Glenn Cohen offered a template for public action to remove the ban on mitochondrial replacement techniques by creating a coalition of patients, families, advocacy groups, physicians, and "sympathetic legislators" to push for congressional hearings on the need to prevent mitochondrial diseases.[59] Their multistep approach reminded us of the way in which the British research community emphasized the potential of IVF to prevent the transmission of genetic diseases in its efforts to create support for IVF in the 1980s. Their decision to do so marked an important first step in gaining support for both the technology itself and for regulating it. If efforts such as the one proposed by Adashi and Cohen succeed, could they lead to larger measures, as they did in Britain? We realize that only a significant shift in the political climate will enable Congress to consider these questions. We hope that shift occurs before it is too late.

APPENDIX

ASSISTED REPRODUCTIVE TECHNOLOGIES BY (SOME OF) THE NUMBERS

Figure A.1. Assisted reproductive technology use by age.

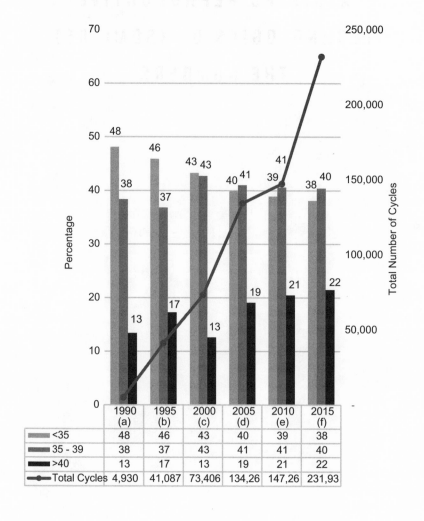

	1990 (a)	1995 (b)	2000 (c)	2005 (d)	2010 (e)	2015 (f)
<35	48	46	43	40	39	38
35 - 39	38	37	43	41	41	40
>40	13	17	13	19	21	22
Total Cycles	4,930	41,087	73,406	134,26	147,26	231,93

(a) **1990**. *Source*: "In Vitro Fertilization-Embryo Transfer (IVF-ET) in the United States: 1990 Results from the IVF-ET Registry," *Fertility and Sterility* 57, no. 1 (1992): 15–23.

(b) **1995**. *Source*: "Assisted Reproductive Technology in the United States and Canada: 1995 Results Generated from the American Society for Reproductive Medicine / Society for Assisted Reproductive Technology Registry," *Fertility and Sterility* 69, no. 3 (1998): 389–98.

(c) **2000**. *Source*: "Assisted Reproductive Technology in the United States: 2000 Results Generated from the American Society of Reproductive Medicine / Society for Assisted Reproductive Technology Registry," *Fertility and Sterility* 81, no. 5 (2004): 1207–20.

(d) **2005**. *Source:* Centers for Disease Control and Prevention, *2005 Assisted Reproductive Technology Success Rates: National Summary and Fertility Clinic Report* (Atlanta, GA: Centers for Disease Control and Prevention, October 2007).

(e) **2010**. *Source:* Centers for Disease Control and Prevention, *2010 Assisted Reproductive Technology Fertility Clinic Success Rates Report* (Atlanta, GA: Centers for Disease Control and Prevention, December 2012).

(f) **2015**. *Source:* Centers for Disease Control and Prevention, *2015 Assisted Reproductive Technology Fertility Clinic Success Rates Report* (Atlanta, GA: Centers for Disease Control and Prevention, December 2017).

Figure A.2. Pregnancy and birth.

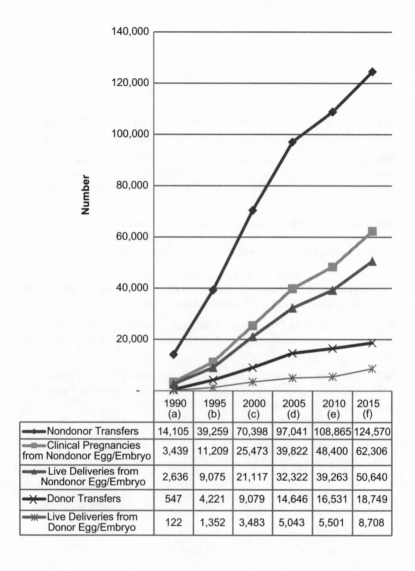

	1990 (a)	1995 (b)	2000 (c)	2005 (d)	2010 (e)	2015 (f)
Nondonor Transfers	14,105	39,259	70,398	97,041	108,865	124,570
Clinical Pregnancies from Nondonor Egg/Embryo	3,439	11,209	25,473	39,822	48,400	62,306
Live Deliveries from Nondonor Egg/Embryo	2,636	9,075	21,117	32,322	39,263	50,640
Donor Transfers	547	4,221	9,079	14,646	16,531	18,749
Live Deliveries from Donor Egg/Embryo	122	1,352	3,483	5,043	5,501	8,708

(a) **1990.** Data taken from assisted reproductive technology treatments initiated from January 1, 1990, to December 31, 1990. Deliveries were counted through October 1991. *Source:* "In Vitro Fertilization-Embryo Transfer (IVF-ET) in the United States: 1990 Results from the IVF-ET Registry," *Fertility and Sterility* 57, no. 1 (1992): 15–23.

(b) **1995.** *Source:* "Assisted Reproductive Technology in the United States and Canada: 1995 Results Generated from the American Society for Reproductive Medicine / Society for Assisted Reproductive Technology Registry," *Fertility and Sterility* 69, no. 3 (1998): 389–98.

(c) **2000.** *Source:* "Assisted Reproductive Technology in the United States: 2000 Results Generated from the American Society of Reproductive Medicine / Society for Assisted Reproductive Technology Registry," *Fertility and Sterility* 81, no. 5 (2004): 1207–20.

(d) **2005.** *Source:* Centers for Disease Control and Prevention, *2005 Assisted Reproductive Technology Success Rates: National Summary and Fertility Clinic Report* (Atlanta, GA: Centers for Disease Control and Prevention, October 2007).

(e) **2010.** *Source:* Centers for Disease Control and Prevention, *2010 Assisted Reproductive Technology Fertility Clinic Success Rates Report* (Atlanta, GA: Centers for Disease Control and Prevention, December 2012).

(f) **2015.** *Source:* Centers for Disease Control and Prevention, *2015 Assisted Reproductive Technology Fertility Clinic Success Rates Report* (Atlanta, GA: Centers for Disease Control and Prevention, December 2017). Note that the total number of births listed for 2015 represents the "number of live births from cycles performed with the intent to transfer at least one embryo. A total of 4,003 cycles were reported with the intent to thaw a previously frozen egg, fertilize the egg, and then transfer the resulting embryo. However because the cycle type (a frozen egg cycle) does not contribute to the calculation of any success rates for the 464 clinics in the 2015 Reproductive Technology Fertility Clinics Success Rates Report, the 4,003 frozen egg cycles are not included in the majority of [the] national report."

Note: For 2005, 2010, and 2015, the numbers were calculated from percentages given. Numbers for these years are close estimates. Given the limited data provided in the sources, it was not possible to separate donor eggs from donor embryos.

Figure A.3. **Multiple births.**

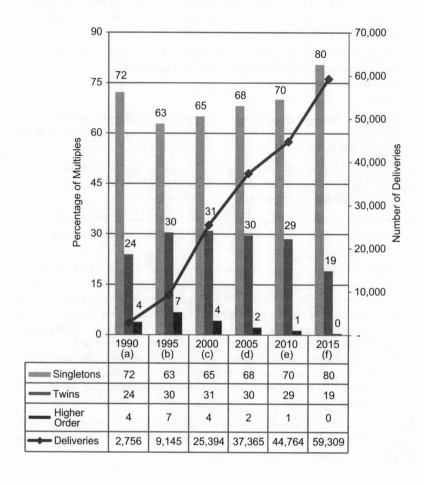

	1990 (a)	1995 (b)	2000 (c)	2005 (d)	2010 (e)	2015 (f)
Singletons	72	63	65	68	70	80
Twins	24	30	31	30	29	19
Higher Order	4	7	4	2	1	0
Deliveries	2,756	9,145	25,394	37,365	44,764	59,309

Note: The numbers in this chart reflect estimates based on best available data from different reports from 1992 to 2017. Some percentages may not add up to 100 percent owing to rounding.

(a) **1990.** Number of deliveries includes all IVF cycles, frozen embryo transfers (ETs), and donor oocytes. Excludes gamete intrafallopian transfer (GIFT) and zygote intrafallopian transfer (ZIFT). *Source:* "In Vitro Fertilization--Embryo Transfer (IVF-ET) in the United States: 1990 Results from the IVF-ET Registry," *Fertility and Sterility* 57, no. 1 (1992): 15–23.

(b) **1995.** Number of deliveries includes all IVF cycles and donor oocytes. No data on multiples were given for frozen ET cycles, so those deliveries were not included in this chart for this year. Excludes GIFT and ZIFT. *Source:* "Assisted Reproductive Technology in the United States and Canada: 1995 Results Generated from the American Society for Reproductive Medicine / Society for Assisted Reproductive Technology Registry," *Fertility and Sterility* 69, no. 3 (1998): 389–98.

(c) **2000.** Number of deliveries includes all ART procedures. Note that, according to the report, "The number of deliveries was equal to the sum of live birth cycles plus stillbirth cycles, which was the same as the sum of cycles that resulted in one or more live born neonates plus cycles that resulted in all stillborn neonates." *Source:* "Assisted Reproductive Technology in the United States: 2000 Results Generated from the American Society of Reproductive Medicine / Society for Assisted Reproductive Technology Registry," *Fertility and Sterility* 81, no. 5 (2004): 1207–20.

(d) **2005.** Number of deliveries includes fresh nondonor cycles, frozen nondonor cycles, and fresh embryos from donor eggs. Data on multiple birth from frozen embryos from donor eggs were not given and have been excluded. The total number of deliveries from all ART procedures with cycles started in 2005 was 38,910; therefore this graph reflects only 96 percent of all deliveries. *Source:* Centers for Disease Control and Prevention, *2005 Assisted Reproductive Technology Success Rates: National Summary and Fertility Clinic Report* (Atlanta, GA: Centers for Disease Control and Prevention, October 2007).

(e) **2010.** Number of deliveries includes fresh nondonor cycles, frozen nondonor cycles, and fresh embryos from donor eggs. Data on multiple birth from frozen embryos from donor eggs were not given and have been excluded. The total number of deliveries from all ART procedures with cycles started in 2010 was 47,090; therefore this graph reflects only 93 percent of all deliveries. *Source:* Centers for Disease Control and Prevention, *2010 Assisted Reproductive Technology Fertility Clinic Report* (Atlanta, GA: Centers for Disease Control and Prevention, December 2012).

(f) **2015.** Number of deliveries includes fresh nondonor cycles, frozen nondonor cycles, fresh embryos from donor eggs, and frozen embryos from donor cycles. The total number of deliveries from all ART procedures with cycles started in 2015 was 60,778. In all, 4,003 cycles were reported with "an intent to thaw a previously frozen egg, fertilize egg, and then transfer the resulting embryo." These cycles were excluded from most of the data in the report and may not be reflected in the graph or number of deliveries presented here. *Source:* Centers for Disease Control and Prevention, *2015 Assisted Reproductive Technology Success Rates* (Atlanta, GA: US Department of Health and Human Services and Centers for Disease Control and Prevention, December 2017).

Table A.1. Infertility Insurance Mandates by States

State (Date Enacted)	Mandate to Cover	Mandate to Offer	Includes IVF Coverage	Excludes IVF Coverage	IVF Coverage Only	Includes Cryopreservation/ Fertility Preservation	Donor Covered	Maximum Limitations	HMO Only
Arkansas[1] (1987)	yes				yes	yes		$15,000	
California (1989)		yes[2]		yes[3]					
Connecticut (1989)	yes[5]		yes			yes	yes		
Delaware (2018)			yes			yes[4] (2017)	yes		
Hawaii (1987)	yes		yes		yes				
Illinois (1991)	yes		yes						
Louisiana (2001)									
Maryland (1985)	yes		yes		yes			$100,000	
Massachusetts (1987; amended 2010)	yes		yes			yes	yes		
Montana (1987)	yes								yes
New Jersey (2001)	yes[6]		yes			yes	yes		
New York (1990, 2002)				yes[7]					
Ohio (1991)	yes								yes
Rhode Island (1989)	yes[8]		yes					$100,000	
Texas (1987)		yes	yes						
West Virginia (1995)	yes								yes

Source: "Infertility Coverage by State," RESOLVE, accessed October 18, 2018, https://resolve.org/what-are-my-options/insurance-coverage/infertility-coverage-state/.

1 A patient's eggs must be fertilized with her spouse's sperm. HMOs are exempted.

2 Insurers are only required to offer services. Employers decide if they want to provide benefits for services: diagnosis, diagnostic testing, medication, surgery, and gamete intrafallopian transfer (GIFT).

3 Only includes GIFT.

4 Mandates insurance coverage of fertility preservation for patients diagnosed with cancer, or whenever deemed medically necessary.

5 IVF retrievals must be completed before a patient is forty-five years of age and transfers completed before a patient is fifty.

6 Experimental or investigational treatments are not covered. Patients must be under forty-six years of age.

7 Provides coverage for diagnosable and correctible medical conditions. Coverage applies to patients twenty-one to forty-four years of age.

8 Coverage applies to women twenty-five to forty-two years of age.

ACKNOWLEDGMENTS

While researching and writing this book over the past several years, we have been fortunate to receive a great deal of financial, intellectual, and emotional support. We are extremely grateful to the Robert Wood Johnson Foundation for a generous multiyear Investigator Award in Health Policy Research, which provided both of us with dedicated time to devote to this project and the means to hire talented research assistants. We are deeply indebted to the RWJF and to the National Program Office at Boston University, where Alan Cohen and Jed Horwitt encouraged and assisted us in ways large and small. We also bonded with our fellow Class of 2013 Investigator Awardees at our annual investigator meetings and learned a great deal from them: Helena Hanson, Laura Hirshbein, Jean Rhodes, Lainie Ross, Joan Teno, Richard Thistelthwaite, Mary Waters, and Ed Yelin.

We began to do some preliminary thinking about this project when Margaret retired as executive dean of arts and sciences at Rutgers University–Camden in 2011 and joined the faculty. Our research assistants early on were Megan Filoon, Erin Hoesly, and Deborah Valentine. Megan was a law student with excellent legal research skills. Erin and Deb were graduate students in childhood studies. They tracked down sources, organized material, and provided thoughtful insights on the research. For the past four years, we have been extremely fortunate to have Matthew Prickett, a PhD student in childhood studies. We benefited greatly from Matt's superb research skills, insight into the material, and mastery of statistical analysis. Matt is responsible for developing the data for, and creating, the appendix table and figures. He was not just a terrific researcher, however; he is a lot of fun to have around. We were sorry to have to tell him that we did not think we could use his suggested title for the book, "Make Womb for Baby," but we promise that if we ever write an article on uterus transplants or artificial uteruses, we will definitely use it and credit him.

In 2013, Margaret keynoted an international conference on Infertility in History, Science, and Culture, held at the University of Edinburgh, where she introduced some of our ideas for the book. She would like to thank Gayle Davis and Tracey Loughran for organizing the conference and providing such a stimulating intellectual environment. At one point in our research, wanting to gain a better understanding of the history of IVF in Great Britain, Margaret wrote to Martin Hume Johnson, of Cambridge University, who was one of the first two PhD students of Robert Edwards. Professor Johnson quickly responded, sharing his insights, publications, and transcripts of several interviews related to the topic. He was extraordinarily generous to us, people he never even met and we are grateful for that. We also want to thank Jacqueline Wolf and Rene Almeling for excellent advice at various stages of the research, and the anonymous reviewer of our manuscript. Jacqueline Wehmueller, our terrific former editor at Johns Hopkins University Press, read the entire manuscript for us. It was Jackie who urged us to write a book appealing to a broad group of readers. With her advice, we streamlined our prose and our arguments. She is an amazing editor, and we cannot thank her enough. This is our third time publishing a book with Johns Hopkins University Press, and we'd like to thank everyone there who answered questions and offered advice; others worked behind the scenes to produce this and our other two books. We also want to thank our excellent copy editor, Ashleigh McKown.

We are also extremely grateful to the physicians who talked to us, sharing their stories and providing invaluable insights. From IVF pioneers to current leaders in the field, from longtime practitioners of reproductive medicine to specialists in maternal-fetal medicine, all of them were incredibly generous with their time and perspectives: Eli Adashi, Luis Blasco, PonJola Coney, Christos Coutifaris (who gets another thank-you later), Robert Debbs, Alan DeCherney, Esther Eisenberg, Leonore Huppert, Nathan Kase, Richard Marrs, Martin Quigley, Steven Sondheimer, and Edward Wallach. We also want to thank Carmen Rodriguez, assistant to Richard Marrs, and Ettora Burrell, assistant to PonJola Coney, for their kindness in facilitating those conversations.

We owe a special thanks to the librarians and archivists who provided invaluable help to us. First, we want to thank Stacey Peeples, the wonderful archivist at Pennsylvania Hospital, for her support and help in locating documents related to the early days of IVF at the hospital.

We are also grateful to everyone on the staff of the Prints and Photographs Division of the Library of Congress; Jessica Murphy and Scott Podolsky of the Center for the History of Medicine, Countway Library of Medicine, Harvard University; Kate Ugarte and Timothy Wisniewski at the Alan Mason Chesney Medical Archives of the Johns Hopkins Medical Institutions; Jim Thweatt and Christopher Ryland of the History of Medicine Collections and Archives of Vanderbilt University and Darrell Solomon of Vanderbilt's division of communications. For assistance with photographs and permissions, we want to thank Simon Burton and Rachel Holdsworth of Holdsworth Associates; Ralph Drew and Erica Varela at the *Los Angeles Times*; and Vanessa Erlichson at Polaris Images.

Margaret owes major debts of gratitude to her colleagues at the Institute for Health, Health Care Policy, and Aging Research at Rutgers, where she is a faculty member, for helping her to learn about health policy research and providing so much other support. David Mechanic, the founding director, invited her to join the institute in 2012. Its weekly seminars in health policy were invaluable, and so was David's advice and counsel. She is also grateful to her graduate school mentor and longtime friend and colleague, the late Gerald N. Grob; Carol Boyer, the institute's associate director; current director XinQi Dong; and Lucia Schutz and Karen Connaughton, who managed the financial aspects of our RWJF grant so cheerfully and efficiently.

At Rutgers–Camden, both of us want to thank Kate Blair, web designer for the Faculty of Arts and Sciences. She created our website, formatted the figures in the appendix, and worked her magic on several of the illustrations. Margaret is fortunate to have as colleagues two terrific novelists, Lisa Zeidner and Lauren Grodstein, who generously gave their time and talent to helping us, especially with the title for the book, and we are extremely grateful to them. Margaret wants to thank the dean of arts and sciences, Kriste Lindenmeyer, for her steadfast support, and the associate and assistant deans and other members of the dean's office, past and present, for their help in different ways, from managing her accounts to giving her writing advice to providing moral support: Maria Garcia and Iris Rodriguez, for everything, and Maria Buckley, Marlene Druding, Ed Kehoe, Bethany Lawton, Amy Liberi, Howard Marchitello, Andrea Ohrenich, Pennie Prete, Rosa Rivera, Julie Roncinske, Joseph Schiavo, Mary Anne Seville, Louise Waters, and Doreen Wheeler. Two members of the information technology staff have gone above and beyond the call of duty in dealing with our technology issues: Rich Buonpastore and Scott Kuhnel. Margaret is grateful as well to Phoebe Haddon, chancellor of Rutgers University–Camden, and Michael Palis, executive vice chancellor for academic affairs and provost. Barbara Lee, senior vice president for academic affairs at Rutgers University, to whom Margaret reports, has been unfailingly supportive of her work over the years.

Both of us want to express a huge debt of gratitude to Christos Coutifaris, division chief of reproductive endocrinology and infertility at the Perelman School of Medicine of the University of Pennsylvania, for giving very generously of his time and sharing with us his historical files on the early days of IVF at Penn, containing material available nowhere else. Wanda is so grateful to Deborah Driscoll, chair of the Department of Obstetrics and Gynecology at Penn, for her enthusiastic support for this research and for approving Wanda's time away from clinical practice to work on it. At Pennsylvania Hospital, Wanda has been fortunate to work with two wonderful chairs of the Department of Obstetrics and Gynecology. Her former chair, Jack Ludmir, was unfailingly interested in and supportive of the collaboration with Margaret. Her current chair, Steven Ralston, has also been very supportive of her work on this book. Finally, for Wanda, who is a busy clinician with a large roster of patients, the support of her colleagues has helped her to focus on this project on the days we were collaborating. Her amazing nurse, Nicole Sullivan, provided ongoing support in multiple ways, and she and the nursing staff at Penn Medicine Gynecology handled all the patient calls during Wanda's research days. Pam Neff, a wonderful colleague, has been tremendously helpful. Wanda wants to thank all her colleagues at Penn Medicine Fertility Care, and especially Scott Edwards, for answering our many questions about the current practice of ART. Wanda also wants to thank her patients, who have helped shape her understanding of many of the situations discussed in this book.

We both owe enormous debts of gratitude to our spouses. Howard Gillette, Margaret's husband and an urban historian, served as our test reader. His job was to keep us from lapsing into medical jargon or assuming too much knowledge about these technologies on the part of our readers. Wanda's husband, Peter Ronner, a biochemist, ensured that our scientific explanations were both accurate and clear and helped us in other ways large and small. Howard and Peter also gave us a great deal of emotional support and took on extra burdens on the home front. We know we are very lucky in our spouses—the best husbands and brothers-in-law in the world, we are convinced—and we hope they know how grateful we are. Thanks, too, to Wanda's son, Lukas, who is a medical student, for always being willing to help in any way he could, and especially for keeping up with new scientific findings and discoveries related to the project.

So many people, in so many ways, supported us in the research and writing of this book, and we can't thank them enough. If despite of all their terrific advice and counsel, there are errors remaining—well, they are our fault.

NOTES

Introduction. The Past as Prologue

1. Conversation with the authors, May 21, 2018. Unless otherwise noted, Wendy's quotes are from this conversation.

2. Margaret Marsh and Wanda Ronner, *The Empty Cradle: Infertility in America from Colonial Times to the Present* (Baltimore: Johns Hopkins University Press, 1996), 207–9, 221–22.

3. Anita Raghavan, "Some Medical Magic: Test Tube Twins Born at HUP," *Daily Pennsylvanian*, September 12, 1984, 1, 5.

4. Steven Sondheimer, interview with the authors, February 29, 2016.

5. Assisted reproductive technology is defined by the Centers for Disease Control and Prevention and the Society for Assisted Reproductive Technology as "all treatments or procedures that include the handling of human eggs or embryos to help a woman become pregnant." The most common form of ART is in vitro fertilization. Fertility treatments such as intrauterine insemination or surgical procedures to correct underlying medical conditions inhibiting a person's ability to conceive are important fertility treatments, and we discuss them when relevant to our narrative, but they are not ART because they do not involve the handling of eggs or embryos. Centers for Disease Control and Prevention, American Society for Reproductive Medicine, Society for Assisted Reproductive Technology, *2015 Assisted Reproductive Technology Fertility Clinic Success Rates Report* (Atlanta: US Department of Health and Human Services, 2017), 531.

6. Conversation with the authors, May 7, 2018. The quotes are from this interview. The name is a pseudonym, and identifying details have been changed.

7. Women who fail to produce an egg each month may be prescribed Clomid, an oral estrogen receptor modulator, to induce ovulation, which would allow them to conceive in the traditional way. Pergonal, the medication prescribed to Wendy so that her ovaries would mature additional eggs for IVF, was another medication also used to induce ovulation. It was produced by extracting gonadotropins—follicle-stimulating hormone and luteinizing hormone—from the urine of postmenopausal women. These gonadotropins are now produced using recombinant DNA technology, and they have virtually replaced Pergonal. One such drug is Follistim, and there are several others. If the medications fail to produce a pregnancy,

■

the next step for some patients is intrauterine insemination (IUI) using the sperm of the male partner. Ovarian stimulation may also be used with this procedure. For many couples, those interventions—a drug regimen in use for decades with or without IUI, —make it possible for a woman to conceive without having to move on to IVF. Marsh and Ronner, *Empty Cradle*, 66–70, 145.

8. "FAQ: Intracytoplasmic Sperm Injection," UCSF Health, accessed September 25, 2018, https://www.ucsfhealth.org/education/intracytoplasmic_sperm_injection/.

9. Centers for Disease Control and Prevention, *National Public Health Action Plan for the Detection, Prevention, and Management of Infertility* (Atlanta: Centers for Disease Control and Prevention, June 2014), 3.

10. Francine Coeytaux, Marcy Darnovsky, and Susan Berke Fogel, "Assisted Reproduction and Choice in the Biotech Age: Recommendations for a Way Forward," *Contraception Journal* 83, no. 1 (2011): 1–4. Judith Daar, "Federalizing Embryo Transfers: Taming the Wild West of Reproductive Medicine?," *Columbia Journal of Gender and Law* 23, no. 2 (2012): 1–6, calls the idea that reproductive medicine is unregulated "an urban myth" because there is professional regulation and some state tort law.

11. Gail Collins, "The War behind the Abortion War," *New York Times*, April 13, 2011.

12. Adrienne Asch and Rebecca Marmor, "Assisted Reproduction," in *The Hastings Center Bioethics Briefing Book* (Garrison, NY: Hastings Center, 2008), 5–10; Anne Donchin, "In Whose Interest? Policy and Politics in Assisted Reproduction," *Bioethics* 25, no. 2 (2011): 92–101.

13. Ruth Deech and Anna Smajdor, *From IVF to Immortality: Controversy in the Era of Reproductive Technology* (Oxford: Oxford University Press, 2007).

14. Concerns among those committed to a social justice and human rights approach to assisted reproduction and a range of related technologies can be found on the website of the Tarrytown Meetings, which were held in 2010, 2011, and 2012 "to address challenges raised by profoundly consequential human biotechnologies and related emerging technologies." In addition to videos, reports, and some full presentations, there is a section containing a trove of articles and related documents; see http://thetarrytownmeetings.org/.

15. See, for example, Charlotte Alter, Diane Tsai, and Francesca Trianni, "What You Really Need to Know about Egg Freezing," *Time*, July 16, 2015, http://time.com/3959487/egg-freezing-need-to-know/.

16. "Winning in the States," Freedom to Marry, accessed February 11, 2015, http://www.freedomtomarry.org/states/; Adam Liptak, "Supreme Court Ruling Makes Same-Sex Marriage a Right Nationwide," *New York Times*, June 26, 2015.

17. Natalie Angier, "The Baby Boom for Gay Parents," *New York Times*, November 26, 2013, D4.

18. See the Human Fertilisation and Embryology Authority website, https://www.hfea.gov.uk/.

Chapter 1. Test-Tube Babies Just around the Corner

1. Mrs. M. C. to John Rock, August 17, 1944, and John Rock to Mrs. M. C., August 31, 1944, Series II A1, Research Records, 1936–1983, Series 7, Box 22, Folder 43, John Rock Papers, Countway Library of Medicine (JR-CLM). Margaret Marsh and Wanda Ronner, *The Empty Cradle: Infertility in America from Colonial Times to the Present* (Baltimore: Johns Hopkins University Press, 1996), 178–80, also includes quotes from various letters.

2. John Rock and Miriam Menkin, "In Vitro Fertilization and Cleavage of Human Ovarian Eggs," *Science* 100, no. 2588 (August 4, 1944): 105–7. At the time in the United States, for such research to be taken seriously, it was important for it to be announced in *Science*.

3. Clipping by Robert S. Bird, "A Human Ovum Is Fertilized in a Test Tube for the First Time," unidentified but likely from a Boston newspaper from August 1944, Series II A1, Research Records, 1936–1983, Box 3, Folder 54, JR-CLM.

4. See, for example, Mrs. G. P. H. to Miriam Menkin, November 23, 1944, Series II A1, Research Records, 1936–1983, Series 7, Box 22, Folder 43, JR-CLM; Mrs. P. C. H. to Miriam Menkin, November 21, 1944, Series II A1, Research Records, 1936–1983, Series 7, Box 22, Folder 43, JR-CLM; Mrs. H. M. D. to "Dear Sirs," August 17, 1944, in Marsh and Ronner, *Empty Cradle*, 178.

5. N. Sproat Heaney, "A Simple Method of Testing the Patency of the Fallopian Tubes," *Gynecological Transactions* 48 (1923): 218; Marsh and Ronner, *Empty Cradle*, 145.

6. Marsh and Ronner, *Empty Cradle*, 175.

7. John Rock to Mrs. M. C., August 31, 1944; John Rock to Mrs. T. R. M. October 16, 1944; John Rock to Mrs. H. M. D., August 31, 1944, Series II A1, Research Records, 1936–1983, Series 7, Box 22, Folder 43, JR-CLM; John Rock to Mrs. A. I., May 4, 1945, Series II A1, Research Records, 1936–1983, Series 7, Box 22, Folder 27, JR-CLM. Margaret Marsh and Wanda Ronner, *The Fertility Doctor: John Rock and the Reproductive Revolution* (Baltimore: Johns Hopkins University Press, 2008), 132–35.

8. Walter E. Duka and Alan H. DeCherney, *From the Beginning: A History of the American Fertility Society, 1944–1994* (Birmingham, AL: American Fertility Society, 1994), 79.

9. Jason M. Colby, "'Banana Growing and Negro Management': Race, Labor, and Jim Crow Colonialism in Guatemala, 1884–1930," *Diplomatic History* 30, no. 4 (September 2006): 596, 605–6.

10. Marsh and Ronner, *Fertility Doctor*, chap. 1.

11. For more detail, see Marsh and Ronner, *Fertility Doctor,* chaps. 3–5.

12. Residencies were shorter in those years. In all, Rock spent two and a half years in post–medical school training.

13. On the discovery of estrogen and progesterone, see Marsh and Ronner, *Empty Cradle*, 138–42; H. Maurice Goodman, "Discovery of the Luteinizing Hormone of the Anterior Pituitary Gland," *American Journal of Physiology—Endocrinology and Metabolism* 287 (November 2004): 818–19.

14. Marsh and Ronner, *Fertility Doctor*, 70–71. In 1934, Rock and Bartlett began to correlate the days of the menstrual cycle with specific features of the endometrium, which for the first time allowed doctors to determine whether a woman had not *yet* ovulated, if she had *already* ovulated, or if she had *failed* to ovulate in that cycle. Without such knowledge, it would be impossible to treat ovulatory failure, because it would not be possible to know whether a woman suffered from it. These data showed that ovulation occurred roughly fourteen days before the next menstrual cycle, but it was not yet possible to *predict* ovulation.

15. Hertig, "A Fifteen Year Search for First Stage Human Ova," *JAMA* 261, no. 3 (January 20, 1989), 435. This was decades before the availability of ultrasound. See Marsh and Ronner, *Fertility Doctor*, esp. 92–93.

16. We discuss this research in *Fertility Doctor*, 87–99. In the 1980s, there was some retrospective condemnation of Hertig and Rock's work because the embryos were obtained while performing hysterectomies within two weeks of women's presumed fertile periods, and some of these women had clearly conceived. This view is understandable; however, for *The Fertility*

Doctor, Wanda reviewed the patient records for this study, and Margaret reviewed all the staff notes and communications with patients. Wanda concluded that the surgeries were warranted based on the medical knowledge of the era, and that any urgent operations were performed immediately. Staff notes and correspondence provide evidence that the patients were informed of the nature of the research and understood why they were not using birth control during the month in which it was scheduled. There does not appear to have been any pressure to participate, and surviving patient correspondence suggests that the women seemed untroubled by the idea of having intercourse so long as they knew the surgery would occur when scheduled. We cannot know, of course, whether the very fact that they were made to feel important by participating exerted any pressure on them. In this era, women generally did not consider themselves pregnant until they missed at least one, possibly two periods, and there were no tests to show otherwise.

17. Tian Zhu, "In Vitro Fertilization," *The Embryo Project Encyclopedia*, July 22, 2009, http://embryo.asu.edu/handle/10776/1665. See C. R. Austin and Arthur Walton, "Fertilisation," in *Marshall's Physiology of Reproduction*, 3rd ed., vol. 1, Part 2, ed. A. S. Parkes (Longmans Green, 1960), chap. 10, 393–96. A copy of Miriam Menkin's typewritten notes on this chapter are in Margaret Marsh's possession. Her original notes are in the Countway Library: [Miriam Menkin], "In Vitro Fertilization," May 9, 1963.

18. Philip Pauly, *Controlling Life: Jacques Loeb and the Engineering Ideal in Biology* (New York: Oxford University Press, 1987), 93, 101. *Arrowsmith* was awarded the Pulitzer Prize in 1926, a prize that Lewis declined to accept.

19. Pauly, *Controlling Life*, 103; Adele E. Clarke, *Disciplining Reproduction: Modernity, American Life Sciences, and the Problems of Sex* (Berkeley: University of California Press, 1998), 249.

20. The community was Woodbine, a fascinating turn-of-the-century community founded for Jews displaced by pogroms and discrimination in their home villages and financed by Jewish philanthropists. The town's synagogue is now a museum, there is a beautiful old cemetery, and many of the original houses are still standing.

21. Any student at a Catholic school could have told the media that "Immaculate Conception" does not mean conception without sex but conception without original sin, and that it applies to the conception of Mary, the mother of Jesus. Still, the name stuck.

22. Leon Speroff, *A Good Man: Gregory Goodwin Pincus—The Man, His Story, the Birth Control Pill* (Portland, OR: Arnica Press, 2009), 86; William L. Laurence, "Life Is Generated in a Scientist's Tube," *New York Times*, March 27, 1936. Many other papers also picked up the story. Pincus conducted two types of reproductive research on rabbits—in vitro fertilization and parthenogenesis—which his fellow researchers knew were different processes. The press, however, often conflated them.

23. Laurence, "Life Is Generated." Aldous Huxley, *Brave New World* (New York: Harper Perennial Modern Classics, 2006). [Originally published in 1932.]

24. J. D. Ratcliff, "No Father to Guide Them," *Collier's*, March 20, 1937, 19, 73. Quotation on p. 73. This is one of the articles that does not draw a clear enough distinction between Pincus's IVF experiments, which were controversial enough, and his work on parthenogenesis. Clarke, *Disciplining Reproduction*, 249, contrasts the negative reaction to Pincus with the "neutral to miraculous portrayals that had greeted Loeb's similar efforts in an earlier era when the miracles of modern science were unquestioned."

25. Marsh and Ronner, *Fertility Doctor*, 140–46. War in Europe loomed while Pincus was at Cambridge, and as a Jewish academic, he was in a terrible situation. He attempted to return to the United States, but he could not find an academic job. He was finally offered an unsalaried

position at Clark University in Massachusetts, with Nathaniel Lord Rothschild promising to pay his salary temporarily. Within a few years, Pincus and Hudson Hoagland established the Worcester Foundation for Experimental Biology, an independent research institute where he spent the rest of his career. It was there that Pincus developed the oral contraceptive, later bringing on John Rock as his clinical collaborator. Speroff, *Good Man*, 21–38 and 84–92.

26. Editorial, "Conception in a Watch Glass," *New England Journal of Medicine* 217 (October 21, 1937): 678; Miriam Menkin, "Notes for Lecture, American Association of Anatomists," 1948, Typescript, p. 3, Series VII, Miriam Menkin Personal Records, Box 22, Folder 62, JR-CLM.

27. Miriam Menkin, untitled talk on in vitro fertilization at Cold Spring Harbor, on either July 18 or 28 (the first number is illegible), 1949, p. 5, Series III B, Box, 19, Folder 33, JR-CLM.

28. Miriam Menkin and John Rock, "In Vitro Fertilization and Cleavage of Human Ovarian Eggs," *American Journal of Obstetrics and Gynecology* 55, no. 3 (March 1948): 440–51. On Menkin and the difficulties she faced as a woman scientist, see Sarah Rodriguez, "Watching the Watch-Glass: Miriam Menkin and One Woman's Work in Reproductive Science, 1938–1952," *Women's Studies* 44, no. 4 (2015): 451–67.

29. A total of 947 women volunteered. About two-thirds of them either did not need surgery or did not require the removal of any ovarian tissue. About one-third had ovarian tissue removed, but only 47 of those women had retrievable eggs. From the 47 women, Menkin retrieved 138 usable eggs. Marsh and Ronner, *Fertility Doctor*, 104–10.

30. Menkin, talk on in vitro fertilization, p. 5.

31. Menkin, talk on in vitro fertilization, p. 14.

32. John Rock and Miriam Menkin, "In Vitro Fertilization and Cleavage of Human Ovarian Eggs," *Science* 100, no. 2588 (August 4, 1944): 105–7. Publication in *Science* was practically required for research to be taken seriously during this period. The issue of whether Rock and Menkin achieved fertilization can never be settled, of course, but most experts at the time believed that they had done so. Marsh and Ronner, *Fertility Doctor*, 114.

33. George L. Streeter to John Rock, July 12, 1944, Series II, Box 15, Folder 103, JR-CLM.

34. Carl Hartman to John Rock, June 8, 1954, Series II A1, Hartman, C. G. In Vitro Fertilization, Ortho Research File, 1954, Box 3, Folder 55, JR-CLM.

35. Joan Younger, "Life Begins in a Test Tube," *Collier's*, April 10, 1945, 27, 49. Quotation on p. 27.

36. For example, see "Test Tube Babies," *New York Times*, August 6, 1944; Howard W. Blakeslee, "Ova of Humans Fertilized in Test Tube for First Time," *Washington Post*, August 5, 1944, 1. The subtitle was "Not Baby Blueprint."

37. *Time,* August 14, 1944, 75. Blakeslee, "Ova of Humans Fertilized," 1.

38. Robert S. Bird, "A Human Ovum Is Fertilized in Test Tube for the First Time."

39. Younger, "Life Begins in a Test Tube," 27, 49. Quotation on p. 49.

40. Marsh and Ronner, *Empty Cradle*, 171–86.

41. Younger, "Life Begins in a Test Tube," 27.

42. For examples, see Marsh and Ronner, *Empty Cradle*, introduction and chaps. 3 and 4.

43. Transcript from tape of Meeting of the Board of Trustees of the Rock Reproductive Study Center on Wednesday, October 23, 1957, Typescript, p. 17, Series 1B, Administrative Records, Box 37, Folder 15, JR-CLM.

44. See Marsh and Ronner, *Fertility Doctor*, 99–103, 123–27.

45. Luigi Mastroianni, interview with Wanda Ronner, July 5, 2007. There are no written consent forms from that era. Marsh and Ronner, *Fertility Doctor*, 108.

46. Younger, "Life Begins in a Test Tube," 27.

47. Letter from Mrs. T. R. M. to Miriam Menkin, October 9, 1944, and Rock's response, October 16, 1944, Series 7, Box 43, JR-CLM. Many letters went to Rock, but a significant number came to Menkin. Because she was not a physician, she passed them along to Rock to answer them.

48. *Time,* August 14, 1944, 75.

49. John Rock to Mrs. N. W., November 29, 1945, Series 7, Box 22, Folder 27, JR-CLM.

50. John Rock to Mrs. J. P., July 30, 1948, Series 7, Box 22, Folder 27, JR-CLM.

51. J. D. Ratcliff, "Babies by Proxy," *Look,* January 31, 1950, 44.

52. John Rock to Mrs. E. M., February 5, 1950, Series 7, Box 22, Folder 27, JR-CLM.

53. John Rock to Mr. B. L., June 22, 1951, Series 7, Box 22, Folder 32, JR-CLM.

54. Marsh and Ronner, *Empty Cradle,* 194; Marsh and Ronner, *Fertility Doctor,* 135 and chap. 6.

55. Loretta McLaughlin, *John Rock, the Pill, and the Church: The Biography of a Revolution* (Boston: Little, Brown, 1982), 87.

56. Landrum B. Shettles, "A Morula Stage of Human Ovum Developed in Vitro," *Fertility and Sterility* 6, no. 4 (1955): 287–89.

57. "Scientists Grow a Human Embryo," *New York Times,* January 14, 1961, 21.

58. Interview with Daniele Petrucci, *CBC Television News,* aired February 9, 1961, on CBC, http://www.cbc.ca/archives/categories/health/reproductive-issues/fighting-infertility/test -tube-baby-experiments.html. In 1964, there were reports that Petrucci claimed to have overseen twenty-eight IVF births, but these births were never verified. *The Ottawa Citizen,* September 24, 1964. See also Marsh and Ronner, *Empty Cradle,* 230.

59. Jane Brody, "Test Tube Fertilization," *New York Times,* March 6, 1966, E6.

Chapter 2. From First Dream to First Baby

1. Robert Edwards and Patrick Steptoe, *A Matter of Life: The Story of a Medical Breakthrough* (New York: William Morrow, 1980), 179.

2. Edwards and Steptoe, *Matter of Life,* 166.

3. Robert Edwards, *Life before Birth* (New York: Basic Books, 1989), 7–8; "The Test-Tube Baby," *Newsweek* 92 (July 24, 1978): 76; "The First Test-Tube Baby," *Time* 112 (July 31, 1978): 58; "Louise: Birth of a New Technology," *Science News* 114 (August 5, 1978): 84; "Test-Tube Baby: It's a Girl," *Time* 112 (August 7, 1978): 68; "All about that Baby," *Newsweek* 92 (August 7, 1978): 66.

4. Martin H. Johnson, "Robert Edwards: The Path to IVF," *Reproductive BioMedicine Online* 23 (2011): 255–56.

5. Robert G. Edwards, "Patrick Christopher Steptoe, C.B.E. 9 June 1913—22 March 1988," *Biographical Memoirs of Fellows of the Royal Society* 42 (1996): 432–52.

6. Walter Sullivan, "Dr. Patrick Steptoe Is Dead at 74; Opened Era of 'Test Tube' Babies," *New York Times,* March 23, 1988, D27.

7. Member, Royal College of Obstetricians and Gynaecologists is MRCOG. Fellow of the Royal College of Surgeons is FRCS. In England, a consultant physician is a highly trained specialist with a listing on a specialist registry. These highly coveted jobs were extremely competitive.

8. Edwards and Steptoe, *Matter of Life,* 60.

9. Edwards, "Patrick Christopher Steptoe," 437.

10. Grzegorz S. Litynski, "Raoul Palmer, World War II, and Transabdominal Coelioscopy: Laparoscopy Extends into Gynecology," *Journal of the Society of Laparoendoscopic Surgeons* 1, no. 3 (July–September 1997): 289–92, https://www.ncbi.nlm.nih.gov/pmc/articles/PMC301 6739/.

11. Grzegorz S. Litynski, "Patrick Steptoe: Laparoscopy, Sterilization, the Test-Tube Baby, and Mass Media," *Journal of the Society of Laparoendoscopic Surgeons* 2 (1998): 99–101; Edwards, "Patrick Christopher Steptoe," 440. Johnson, "Robert Edwards," 254.

12. Palmer had himself used the laparoscope to view the ovary, as described in Litynski, "Raoul Palmer," 289–92.

13. Much of this discussion of Edwards is drawn from Johnson, "Robert Edwards," 245–62.

14. A council house is a form of public housing. Johnson, "Robert Edwards," 246.

15. Johnson, "Robert Edwards," 246.

16. Johnson, "Robert Edwards," 246.

17. Johnson, "Robert Edwards," 247.

18. Edwards and Steptoe, *Matter of Life*, 17.

19. Edwards and Steptoe, *Matter of Life*, 20.

20. Johnson, "Robert Edwards," 248.

21. Johnson, "Robert Edwards," 248–50; Edwards and Steptoe, *Matter of Life*, 39–40.

22. Alan Beatty, Edwards's mentor, believed that Pincus's work on parthenogenesis in rabbits was accurate, as did Leon Speroff, a distinguished twentieth-century reproductive endocrinologist and Pincus biographer. Leon Speroff, *A Good Man: Gregory Goodwin Pincus, the Man, His Story, the Birth Control Pill* (Portland, OR: Arnica, 2009), 89–91, 94.

23. Edwards and Steptoe, *Matter of Life*, 43.

24. "Men in the News: Mr Steptoe and Dr Edwards," *The Times*, July 28, 1978, 5.

25. R. G. Edwards, B. D. Bavister, and P. C. Steptoe, "Early Stages of Fertilization In Vitro of Human Oocytes Matured In Vitro," *Nature* 221 (February 15, 1969): 632–35.

26. W. J. Hamilton and T. W. Glenister, "Human Life in the Test Tube," Letters to the Editor, *The Times*, February 19, 1969, 9; *Nature* 221 (March 8, 1969): 981.

27. Edwards and Steptoe, *Matter of Life*, 85–86.

28. R. G. Edwards, P. C. Steptoe, and J. M. Purdy, "Fertilization and Cleavage In Vitro of Preovulator Human Oocytes," *Nature* 227 (September 26, 1970): 1307–09; P. C. Steptoe, R. G. Edwards, and J. M. Purdy, "Human Blastocysts Grown in Culture," *Nature* 229 (January 8, 1971): 132–33.

29. Joan Arehart-Treichel, "Test-Tube Babies in the Making," *Science News* 103, no. 8 (February 24, 1973): 125.

30. Jean Marx, "Out of the Womb—Into the Test Tube," *Science*, n.s., 182, no. 4114 (November 23, 1973): 814.

31. Robert Edwards, *Life before Birth*, 7, 168–69.

32. Edwards and Steptoe, *Matter of Life*, 106–7.

33. Edwards and Steptoe, *Matter of Life*, 108–9.

34. Arehart-Treichel, "Test-Tube Babies in the Making," 126.

35. Johnson, "Robert Edwards," 258.

36. John Leeton, *Test Tube Revolution: The Early History of IVF* (Clayton, VIC: Monash University, 2013), ix, 14.

37. Robin Marantz Henig, *Pandora's Baby: How the First Test Tube Babies Sparked the Reproductive Revolution* (New York: Houghton Mifflin, 2004), 31–32, 273.

38. "'Test Tube' Baby Alive and Well in Britain," *The Times*, July 16, 1974, 1; "Call for Details about First Test Tube Baby, *The Times*, July 17, 1974, 2.

39. "Test-Tube Babies: Now a Reality?," *Science News*, July 20, 1974, 106; "The Baby Maker," *Time* 104 (July 30, 1974): 58; "Test Tube Babies: Reaction Sets In," *Science News* 106 (July 27, 1974): 53; "Test-Tube Babies?," *Newsweek* 84 (July 29, 1974): 70; Henig, *Pandora's Baby*, 114.

40. Landrum B. Shettles, "A Morula Stage of Human Ovum Developed In Vitro," *Fertility and Sterility* 6, no. 4 (1955): 287–89.

41. Henig, *Pandora's Baby*, 20–21. Quotation on 21.

42. "Test Tube Bereavement," *Newsweek* 92 (July 31, 1978): 70.

43. Henig, *Pandora's Baby*, 103.

44. "Motherhood—Who Needs It?," *Look* 34 (September 22, 1970): 17; "Make Love, Not Babies," *Newsweek* 75 (June 15, 1970): 111. Anti-child articles became pervasive in the 1970s. See "Childless Bliss," *Newsweek* 84 (December 9, 1974): 87; "If You Had It to Do Over Again—Would You Have Children?," *Good Housekeeping* 182 (June 1976): 100–101ff.; Tilla Vahanian and Sally Wendkos Olds, "Will Your Children Break . . . Or Make . . . Your Marriage?," *Parents* 49 (August 1974): 79; J. E Veevers, "Voluntary Childlessness: A Review of Issues and Evidence," *Marriage and Family Review* 2, no. 2 (Summer 1979): esp. 11–14.

45. "Down with Kids," *Time* 100 (July 3, 1972): 35; "Kidding You Not," *Newsweek* 82 (November 5, 1973): 82; "Childless Bliss," 87; "Those Missing Babies," *Time* 104 (September 1974): 54. The best assessment of NON is by Jenna Healey, "Rejecting Reproduction: The National Organization for Non-Parents and Childfree Activism in 1970s America," *Journal of Women's History* 28, no. 1 (Spring 2016): 131–56.

46. William D. Mosher and William F. Pratt, "Fertility and Infertility in the United States," *Advance Data* 192 (December 4, 1990): 3–5.

47. Margaret Marsh and Wanda Ronner, *The Empty Cradle: Infertility in America from Colonial Times to the Present* (Baltimore: Johns Hopkins University Press, 1996), 210–12, 214–15, 245.

48. National Center for Health Statistics, *Vital Statistics of the United States, 1980*, vol. 1, *Natality*, DHHS Pub. No. (PHS) 85-100 (Washington, DC: Government Printing Office, 1984), 1–7.

49. Boston Women's Health Collective, *Ourselves and Our Children* (New York: Random House, 1978), 17–32.

50. Elizabeth Siegel Watkins, *The Estrogen Elixir: A History of Hormone Replacement Therapy in America* (Baltimore: Johns Hopkins University Press, 2007), 132.

51. Boston Women's Health Collective, *Ourselves and Our Children*, 26–27; Shulamith Firestone, *The Dialectic of Sex: The Case for Feminist Revolution* (New York: William Morrow), 11, 197–9.

52. Leon Kass quote from Department of Health, Education and Welfare, Ethics Advisory Board, *Report and Conclusions: HEW Support of Research Involving Human In Vitro Fertilization and Embryo Transfer*, May 4, 1979, Reprinted in Clifford Grobstein, *From Chance to Purpose: An Appraisal of External Human Fertilization* (Reading, MA: Addison-Wesley, 1981), quote on p. 182. Leon Kass, "Babies by Means of In Vitro Fertilization: Unethical Experiments on the Unborn?," *New England Journal of Medicine* 285, no. 21 (November 18, 1971): 1174–79.

53. Edwards and Steptoe, *Matter of Life*, 113.

54. Edwards and Steptoe, *Matter of Life*, 113.

55. Edwards and Steptoe, *Matter of Life*, 114–15. "Storm over Work on Test Tube Babies," *The Times*, October 18, 1971, 5.

56. Georgeanna Jones, "Women—The Impact of Advances in Fertility Control on Their Future: A Presidential Address," *Fertility and Sterility* 22, no. 6 (June 1971): esp. 347, 349.

57. Edwards and Steptoe, *Matter of Life*, 108.

58. "In Vitro Fertilization of Human Ova and Blastocyst Transfer: An Invitational Symposium," *Journal of Reproductive Medicine* 11, no. 5 (November 1973): 192–94, 201–2.

59. Jones, "Women," 348.

60. Grobstein, *From Chance to Purpose.*

61. "Fertilization outside Womb," *Science Digest* 69 (January 1971): 90.

62. Edwards and Steptoe, *Matter of Life*, 121.

63. Leeton, *Test Tube Revolution*, 14.

64. Edwards and Steptoe, *Matter of Life*, 138–39.

65. "Test-Tube Bereavement," *Newsweek* 92 (July 31, 1978): 70; Doris Del Zio as told to Suzanne Wilding, "I Was Cheated of My Test-Tube Baby," *Good Housekeeping* 188 (March 1979): 202; "Detour on the Road to Brave New World," *Science* 210 (August 4, 1978): 424–25.

66. "Test-Tube Baby," *Newsweek*, 76.

67. Linda Witt, "Two Nashville Doctors May Help an American Mother Have Her Own Test-Tube Baby," *People* 10, no. 7 (August 14, 1978); Soupart died in 1981. In 1972, when Soupart provided visual evidence of fertilization using an electron microscope, some argued that his work provided more definitive evidence than that of Steptoe and Edwards. "Pierre Soupart, Pioneer in Laboratory Fertilization," *New York Times*, June 12, 1981; Marie-Claire Orgebin-Crist, "Pierre Soupart, 1923–1981," *Journal of Andrology* 3, no. 6 (November–December 1982): 354.

68. Witt, "Two Nashville Doctors."

69. "A Young Couple Await Their Test-Tube Baby," *Ebony* 34, no. 1 (November 1978): 33–39; Marsh and Ronner, *Empty Cradle*, 238–39.

70. "Test-Tube Baby: It's a Girl," *Time* 112 (August 7, 1978): 68.

Chapter 3. IVF Comes to America

1. Howard W. Jones Jr. and Georgeanna Seegar Jones, *War and Love: A Surgeon's Memoir of Battlefield Medicine with Letters to and from Home* (Bloomington, IN: XLibris, 2004), 186.

2. *In-Vitro Fertilization Oversight: Hearing before the Subcommittee on Health and the Environment of the Committee on Interstate and Foreign Commerce*, 95th Cong., second session, 111 (August 4, 1978); Clifford Grobstein, *From Chance to Purpose: An Appraisal of External Human Fertilization* (Reading, MA: Addison-Wesley, 1981), 155–59.

3. Tanya Melich, *The Republican War against Women: An Insider's Report from Behind the Lines* (New York: Bantam Books, 1996), 60, 77, 89–92.

4. "1976 Democratic Party Platform," July 12, 1976, American Presidency Project, http://www.presidency.ucsb.edu/ws/?pid=29606.

5. For the larger context, see Robert Self, *All in the Family: The Realignment of American Democracy since the 1960s* (New York: Hill and Wang, 2012), 256, 310, 321, and 371.

6. Harry Minium and Debbie Messina, "Remembering Norfolk's Visionary, Dr. Mason Andrews," *Virginia Pilot*, October 14, 2006; Patricia Sullivan, "Mason Andrews: In Vitro Pioneer Physician," *Washington Post*, October 15, 2006.

7. Victor Cohn, "Nation's First 'Test-Tube' Baby Due within Days in Program at Norfolk," *Washington Post*, December 25, 1981, A6.

8. Randi Hutter Epstein, "Howard W. Jones, Jr., a Pioneer of Reproductive Medicine, Dies at 104," *New York Times*, July 31, 2015, http://www.nytimes.com/2015/08/01/science/howard-w-jones-jr-a-pioneer-of-reproductive-medicine-dies-at-104.html?_r=0.

9. "Celebrating Women Physicians," Changing the Face of Medicine, accessed December 2, 2016, https://cfmedicine.nlm.nih.gov/physicians/biography_291.html.

10. Walter E. Duka and Alan H. DeCherney, *From the Beginning: A History of the American Fertility Society, 1944–1994* (Birmingham, AL: American Fertility Society, 1994), 118.

11. Jones and Jones, *War and Love*, 547.

12. "Georgeanna Seegar Jones, Pioneer in Reproductive Medicine," National Library of Medicine, accessed November 15, 2018, https://www.nlm.nih.gov/locallegends/Biographies /Jones_Georgeanna.html; Stephen C. Ruffenach, "Georgeanna Seegar Jones," *Embryo Project Encyclopedia*, July 22, 2009, http://embryo.asu.edu/handle/10776/2002.

13. Jones and Jones, *War and Love*, 550. Today, there is disagreement over whether such surgery should be performed on infants. See Epstein, "Howard W. Jones, Jr."

14. Epstein, "Howard W. Jones, Jr."

15. Robert Edwards and Patrick Steptoe, *A Matter of Life: The Story of a Medical Breakthrough* (New York: William Morrow, 1980), 53–54.

16. Edwards and Steptoe, *Matter of Life*, 53–54.

17. Randi Hutter Epstein, "Pioneer Reflects on Future of Reproductive Medicine," *New York Times*, March 22, 2010; Howard Jones, *In Vitro Fertilization Comes to America: Memoir of a Medical Breakthrough* (Williamsburg, VA: Jamestowne Bookworks, 2014), 7–8; Mark Zhang, "Howard Wilbur Jones Jr," *Embryo Project Encyclopedia*, March 3, 2013, http://embryo.asu.edu/handle/10776/4216.

18. Duka and DeCherney, *From the Beginning*, 164.

19. Glenn Frankel, "Test-Tube Baby Clinic: Approval of Norfolk Hospital Lab Stirs Anger," *Washington Post*, January 9, 1980, C1.

20. "Hearings Asked on Va. Clinic for Test-Tube Babies," *Washington Post*, January 15, 1980, B2; Richard Cohen, "Test-Tube Babies: Why Add to a Surplus?," *Washington Post*, February 3, 1980, B1; Ellen Goodman, "The Baby Louise Clinic," *Washington Post*, January 15, 1980, A15.

21. Editorial, *Washington Post*, January 19, 1980, A14.

22. Glenn Frankel, "Test-Tube Baby Clinic Wins Round," *Washington Post*, February 13, 1980, C1.

23. Anne Taylor Fleming, "New Frontiers in Conception," *New York Times Magazine*, July 20, 1980, 49.

24. John Leeton, *Test Tube Revolution: The Early History of IVF* (Melbourne: Monash University, 2013), 31–32

25. Leeton, *Test Tube Revolution*, 29.

26. Cohn, "Nation's First," A6–A7; "'Test-Tube' Baby Born in U.S., Joining Successes Around World," *New York Times*, December 29, 1981, 1, C1.

27. Sandra Bookman and Dale Russakoff, "'It's Positive'—How Norfolk School Created Test-Tube Pregnancy," *Washington Post*, May 17, 1981, B1.

28. Victor Cohn, "HEW Urged to Lift Ban on Test Tube Birth Studies," *Washington Post*, September 16, 1978, A2.

29. Victor Cohn, "Test Tube Baby Study Wins Approval," *Washington Post*, February 4, 1979, A5.

30. Clifford Grobstein, Michael Flower, and John Mendeloff, "External Human Fertiliza-

tion: An Evaluation of Policy," *Science*, n.s., 222, no. 4620 (October 14, 1983): 131, quotes the recommendations of the Ethics Advisory Board.

31. Cohn, "Test Tube Baby Study."

32. Cohn, "HEW Urged to Lift Ban."

33. Advertisement, *New York Times*, March 11, 1979, E5.

34. Ronald M. Green, *The Human Embryo Research Debates: Bioethics in the Vortex of Controversy* (Oxford: Oxford University Press, 2001), 2.

35. Joseph Califano, *Inside: A Public and Private Life* (New York: Public Affairs, 2004), 351, 60–61.

36. Grobstein, *From Chance to Purpose*, 113–14.

37. Grobstein, *From Chance to Purpose*; Stephen S. Hall, *Merchants of Immortality* (New York: Houghton Mifflin Harcourt, 2003), 99.

38. Grobstein, *From Chance to Purpose*, 114.

39. Alan DeCherney, interview with the authors, November 2, 2015.

40. Martin Quigley, interview with the authors, June 9, 2017.

41. Walter Sullivan, " 'Test Tube' Baby Born in U.S."

42. Lawrence K. Altman, "Dr. Michael E. DeBakey, Rebuilder of Hearts, Dies at 99," *New York Times*, July 13, 2008, http://www.nytimes.com/2008/07/13/health/12cnd-debakey.html. Richard Marrs, interview with the authors, July 23, 2017, and October 20, 2017.

43. Marrs, interviews with the authors.

44. Marrs, interviews with the authors.

45. Marrs, interviews with the authors.

46. Marrs, interviews with the authors.

47. John Leeton, *Test Tube Revolution*, 26–28.

48. Marrs, interviews with the authors.

49. Anjani Chandra, Casey E. Copen, and Elizabeth Hervey Stephen, "Infertility and Impaired Fecundity in the United States, 1982–2010: Data from the National Survey of Family Growth," *National Health Statistics Reports* 67 (August 14, 2013): 6.

50. Self, *All in the Family*, 366–83; Melich, *Republican War against Women*, 170–73.

51. Melich, *Republican War against Women*, 142–44; Self, *All in the Family*, 310–11.

52. "Government Urged to Actively Support Test Tube Baby Research," *Washington Post*, January 4, 1982, A3; "U.S. Scientist Barred from Speaking at Workshop," *Washington Post*, September 14, 1982, C3; "Norfolk Team in Forefront of Test-Tube Baby Boom," *Washington Post*, September 13, 1982, A2.

53. Another early program, led by Andrew Silverman at the University of Texas at San Antonio, did not achieve a birth. See "A Team of Doctors Will Institute," UPI Archives, October 24, 1980, http://www.upi.com/Archives/1980/10/24/A-team-of-doctors-early-in-1981-will-institute/8345341208000/; "Vanderbilt University's One-Year-Old 'Test Tube' Baby Clinic Has Produced," UPI Archives, March 2, 1983, http://www.upi.com/Archives/1983/03/02/Vanderbilt-Universitys-one-year-old-test-tube-baby-clinic-has-produced/8693415429200/; "Efforts to Produce a Test-Tube Baby Will Begin," UPI Archives, December 18, 1980, http://www.upi.com/Archives/search/?ss=efforts+to+produce+a+test+tube+baby&s_y=1980&s_m=12&s_d=18&search=1; "The Nation's Third 'Test Tube Baby' a Healthy 7-Pound Girl," UPI Archives, June 9, 1982, http://www.upi.com/Archives/1982/06/09/The-nations-third-test-tube-baby-a-healthy-7-pound/4586392443200/; Martha Woodall, "Creation of 'Test-Tube' Babies Gives Heart to the Infertile," *Philadelphia Inquirer*, December 9, 1982, D1, D9–D11.

54. Gina Kolata, "The Sad Legacy of the Dalkon Shield," *New York Times*, December 6, 1987, http://www.nytimes.com/1987/12/06/magazine/the-sad-legacy-of-the-dalkon-shield.html; Karen Kenney, "Dalkon Shield Gives Birth to a Generation of Lawsuits," *Chicago Tribune*, April 30, 1985, http://articles.chicagotribune.com/1985-04-30/features/8501260779_1_dalkon -shield-intrauterine-robins; Margaret Marsh and Wanda Ronner, *The Fertility Doctor: John Rock and the Reproductive Revolution* (Baltimore: Johns Hopkins University Press, 2008), 275–76.

55. Marsh and Ronner, *Empty Cradle*, 237–38 and 247.

56. Louise Brown, *My Life as the World's First Test-Tube Baby* (Bristol, UK: Bristol Books, 2015), 13–15.

57. Brown, *My Life*, 13–19.

58. John Leeton, *Test Tube Revolution*, 3–4.

59. "Australia's First Test Tube Baby," *Australian Women's Weekly*, July 9, 1980, 2–4. John Leeton, *Test Tube Revolution*, 26–27.

60. Jones, *In Vitro Fertilization Comes to America*, 81–82.

61. Marsh and Ronner, *Empty Cradle*, 142.

62. Marrs, interviews with the authors.

Chapter 4. From Miracle Births to Medical Mainstream

1. Patient story as told to one of the authors; additional information from Christos Coutifaris, interview with the authors, December 16, 2015.

2. Vanderbilt's first stimulation cycle was in February of 1982, and Penn's was in May of that year. *Consumer Protection Issues Involving In Vitro Fertilization Clinics: Hearing before the Subcommittee on Regulation, Business Opportunities, and Energy, of the Committee on Small Business*, 101st Congress, first session, 1055, 1104 (1989). Hereafter Wyden Hearing, 1989.

3. "The Pioneers," *ESHRE Monographs* 4, no. 1 (2005): https://academic.oup.com/eshre monographs/article/2005/4/1/569153/1-The-Pioneers Accessed 2016.

4. Walter Sullivan, "First Test-Tube Baby Born in U.S., Joining Successes around World," *New York Times,* December 28, 1981, 1. Individual clinic data tabulated by Marsh and Ronner with assistance from Matthew Prickett from completed surveys provided to Rep. Ron Wyden. See Wyden Hearing, 1989, raw data on pp. 319–1300; "The Pioneers."

5. Mastroianni had received ongoing NIH funding for years for his IVF research in monkeys; there was apparently some concern at Penn that the agency might cut off his other funding if he engaged in human IVF. Christos Coutifaris told us that it was impossible to say whether this rumor had any basis in fact. Interview with the authors, December 16, 2015. "Government Urged to Actively Support Test-Tube Baby Research," *Washington Post*, January 4, 1982, A3; "U.S. Scientist Barred from Speaking at Workshop," *Washington Post*, September 14, 1982, C3; "Norfolk Team in Forefront of Test-Tube Baby Boom," *Washington Post*, September 13, 1982, A2.

6. "The Pioneers."

7. Jerome F. Strauss III and Luigi Mastroianni Jr., "In Memoriam: Celso-Ramon Garcia, M.D. (1922–2004), Reproductive Medicine Visionary," *Journal of Experimental and Clinical Assisted Reproduction* 2, no. 2 (2005): https://www.ncbi.nlm.nih.gov/pmc/articles/PMC548 289/.

8. Strauss and Mastroianni, "In Memoriam." Before he joined Mastroianni at Penn in 1965, Garcia worked with Gregory Pincus and was in clinical practice with John Rock.

9. "Reported In-Vitro Fertilization of Human Ovum Elicits Doubts," *Ob. Gyn. News*, December 15, 1970, 1, 14–15.

10. Luigi Mastroianni to Daniel Goldberg, January 18, 1982, Typescript. Courtesy of Christos Coutifaris. We have no idea where he got the figure of fifty IVF births.

11. "In Memoriam, 2009: Richard W. Tureck, Obstetrics and Gynecology," Association of Senior and Emeritus Faculty, Perelman School of Medicine, University of Pennsylvania, accessed November 15, 2018, https://www.med.upenn.edu/asef/richard-w.-tureck.html.

12. Luis Blasco, interview with the authors, February 25, 2015.

13. Steven Sondheimer, interview with the authors, February 29, 2016.

14. Sondheimer, interview with the authors; Owen Edmonston, "HUP Physicians Supervise Birth of Test Tube Baby," *Daily Pennsylvanian*, September 26, 1983, 8; Linda Herskowitz, "HUP Program Breeds Success," *Philadelphia Inquirer*, February 24, 1983, 1B–2B.

15. Sondheimer, interview with the authors.

16. Herskowitz, "HUP Program Breeds Success."

17. Martha Woodall, "In Vitro: Creation of 'Test Tube' Babies Gives Heart to the Infertile," *Philadelphia Inquirer*, December 9, 1982, D10. Herskowitz, "HUP Program Breeds Success."

18. Edmonston, "HUP Physicians Supervise Birth," 8. Memo from Ricki Baker to Luigi Mastroianni, "Annual Report," Typescript. Courtesy of Christos Coutifaris.

19. Interviews with the authors: Richard Marrs, June 23, 2017; Alan DeCherney, November 2, 2015; and Edward Wallach, April 6, 2016.

20. Blasco, interview with the authors.

21. In order, the births occurred at Eastern Virginia Medical School (Howard and Georgeanna Jones, December 1981); University of Southern California (Richard Marrs, June 1982); University of Texas Health Science Center-Houston (Martin Quigley, February 1983); Vanderbilt (Anne Colston Wentz, March 1983); and Yale (Alan DeCherney, May 1983).

22. Woodall, "In Vitro," D1, D9–D11. Owen Edmonston, "Test-Tube Baby Is 'Miracle' on 34th Street," *Daily Pennsylvanian*, September 27, 1983, 6–7.

23. Sondheimer, interview with the authors.

24. Esther Eisenberg, interview with the authors, March 14, 2017.

25. Richard Marrs, interview with Margaret Marsh, October 20, 2017.

26. DeCherney, interview with the authors; Sondheimer, interview with the authors.

27. Wallach, interview with the authors.

28. Wallach, interview with the authors.

29. "Turning Hopes into Reality," *Cape Cod Times*, October 21, 1985, 21, 23. Mastroianni spent part of his summers working at the prominent research facility Woods Hole in Cape Cod.

30. Eisenberg, interview with the authors.

31. Wallach, interview with the authors.

32. PonJola Coney, interview with Margaret Marsh, October 13, 2017.

33. Coney, interview with Marsh.

34. Coney, interview with Marsh.

35. Wyden Hearing, 1989. The appendix to his hearing includes 171 questionnaires completed by IVF clinics in 1988. We tabulated and analyzed the data from those questionnaires.

36. Wyden Hearing, 1989.

37. Wyden Hearing, 1989. For 1988, that figure is based on the number of births plus the number of ongoing pregnancies at the end of the year.

38. Many centers used GIFT in the 1980s, but with improvement in IVF success rates and the development of ultrasound-guided egg retrieval, the procedure became virtually obsolete. In 1987 and 1988, Corson's program used GIFT in 251 patients, compared to 49 patients at the University of Pennsylvania. Wyden Hearing, 1989. Statistics from pp. 1055–58 and 1067–70.

39. The program at Albert Einstein Hospital, headed by Martin Freedman, opened in 1985. In 1988, it had fifty patients, two-thirds of them aged thirty-five or younger. Of them, 60 percent had fallopian tube disease and another 24 percent suffered from endometriosis. In 1987, it had a success rate of about 16 percent; counting births and assuming that the ongoing pregnancies at the end of 1988 would end successfully, the success rate was about 12 percent. A new private clinic in the Philadelphia suburb of Melrose Park, co-directed by Jerome Check and Kosrow Nowroozi, opened in 1988. It had 129 patients in its first year of operation. Its demographics trended younger—almost half of its patients were thirty-five and under, and only 6 percent were over forty. Data from Wyden Hearing, 1989.

40. Laurence Cooper, "Creation Science: Everything You Never Want to Know about Sex (until It Doesn't Work)," *Philadelphia Magazine*, November 1988, 157.

41. Coney, interview with Marsh.

42. Susan Lenz and J. G. Lauritsen, "Ultrasonically Guided Percutaneous Aspiration of Human Follicles under Local Anesthesia: A New Method of Collecting Oocytes for In Vitro Fertilization," *Fertility and Sterility* 38, no. 6 (1982): 673–77.

43. Among the doctors we interviewed, the timing of the shift from laparoscopy to ultrasound varied. The earliest was DeCherney, who used transvesical ultrasound as early as 1983, he believes. Others seemed not to have used ultrasound until the transvaginal techniques had been developed—Quigley in 1985, Leonore Huppert sometime after 1986. The program at Pennsylvania Hospital was still using laparoscopy in 1984, and it is not clear when it changed. Coney started using transvaginal ultrasound in 1988, and Marrs recalls using it around 1986 or 1987.

44. P. Dellenbach et al., "Transvaginal Sonographically Controlled Follicle Puncture for Oocyte Retrieval," *Fertility and Sterility* 44 (1985): 656–62; Matts Wikland, Lennart Enk, and Lars Hamberger, "Transvesical and Transvaginal Approaches for the Aspiration of Follicles by Use of Ultrasound," *Annals of the New York Academy of Sciences* 442, no. 1 (1985): 182–94.

45. John Leeton, *Test Tube Revolution: The Early History of IVF* (Melbourne: Monash University, 2013), 49.

46. M. Wikland et al., "Use of a Vaginal Transducer for Oocyte Retrieval in an IVF/ET Program," *Journal of Clinical Ultrasound* 15 (1987): 245–51.

47. This is according to questionnaires completed by the directors of 171 of them. The text of the hearings refers to 146 clinics, but the actual number of received questionnaires is 171. Perhaps the doctors who looked at the surveys eliminated for analytical purposes questionnaires that were not fully completed. Wyden Hearing, 1989.

48. Wyden Hearing, 1989, p. 1256.

49. Wyden Hearing, 1989, pp. 1256–61.

50. Calculated from data from Wyden Hearing, 1989.

51. Calculated from data from Wyden Hearing, 1989.

52. Wyden Hearing, 1989, p. 2.

53. Testimony of Richard Marrs, Wyden Hearing, 1989, p. 33.

54. Testimony of Benjamin Younger, Wyden Hearing, 1989, p. 146.

55. *Consumer Protection Issues Involving In Vitro Fertilization Clinics, Hearing before the Subcommittee on Regulation and Business Opportunities of the Committee on Small Business,* 100th Cong., second session, 1 (1988). Hereafter Wyden Hearing, 1988.

56. Testimony of Alan DeCherney, Wyden Hearing, 1989, p. 136.

57. Testimony of Richard Marrs, Wyden Hearing, 1989, p. 36.

58. Coney, interview with Marsh.

59. Wyden Hearing, 1989. Sampling of programs from the clinic specific data.

60. P. Devroey and A. Van Steirteghem, "A Review of Ten Years Experience of ICSI," *Human Reproduction Update* 10, no. 1 (2004): 19–28. The first procedure was called PZD, for partial zona dissection, and the second was called SUZI, for subzonal insemination.

61. Wyden Hearing, 1989.

62. Today, an egg-sharing program means that two or more women using a donor egg to conceive will use the same donor and split the costs. In the 1980s, it meant that a woman going through IVF and having extra eggs could share those extra eggs with another patient.

63. Leeton, *Test Tube Revolution,* 43–44.

64. Leeton, *Test Tube Revolution,* 44–46.

65. Leeton, *Test Tube Revolution,* 44–47; Peter Lutjen, Alan Trounson, John Leeton, Jock Findlay, Carol Wood, and Peter Renou, "The Establishment and Maintenance of Pregnancy Using In Vitro Fertilization and Embryo Donation in a Patient with Primary Ovarian Failure," *Nature* 307 (January 12, 1984): 174–75.

66. Testimony by Gary Hodgen, *Alternative Reproductive Technologies: Implications for Children and Families, Hearing before the Select Committee on Children, Youth, and Families,* 100th Cong., first session, 6 and 12 (May 21, 1987). Hereafter Morrison Hearing. According to Hodgen, the Jones Institute, the successor to the IVF program created by Howard and Georgeanna Jones and opened in 1980 was formally constituted in 1983–84.

67. Giovanna Breu and Frank Feldinger, "In California, a Small Bundle of Medical History Arrives on Time: The First U.S. Frozen Embryo Baby," *People* 25, no. 25 (June 23, 1986): http://www.people.com/people/archive/article/0,,20093925,00.html.

68. Wyden Hearing, 1989. Ten additional clinics were making plans to do the same, about 8 percent of the total number of IVF clinics at that time, which numbered about 116 to 120. The number of clinics comes from our analysis of the Wyden Hearing data.

69. K. Ames and L. Denworth, "And Donor Makes Three," *Newsweek* 118, no. 14 (September 30, 1991), 60–61. The estimate is for the number of clinics offering the service in 1988 and comes from Jan Hoffman, "Egg Donations Meet a Need and Raise Ethical Questions," *New York Times,* January 8, 1996, 1, 10.

70. A. Sachs and C. Gorman, "And Baby Makes Four," *Time* 136, no. 9 (August 27, 1990): 53.

71. According to the American College of Obstetricians and Gynecologists (ACOG), from 10 to 25 percent of all clinically recognized pregnancies end in miscarriage.

Chapter 5. The Elusive Search for National Consensus

1. Otto Friedrich, "The New Origins of Life," *Time,* September 10, 1984, 50.

2. The *Newsweek* article, quoted everywhere, is hard to find now because the magazine removed it from its website after the retraction. Megan Garber, "When *Newsweek* 'Struck Terror in the Hearts of Single Women,'" *The Atlantic,* June 2, 2016, https://www.theatlantic.com

/entertainment/archive/2016/06/more-likely-to-be-killed-by-a-terrorist-than-to-get-married /485171/. As the *Atlantic* noted, "Thirty years later . . . it's easy to forget that the so-pervasive-as-to-be-Ephroned marriage-and-terrorism stat was plucked from a single piece of journalism that was in turn based on a study that was, at the time of the story's publication, unpublished [and] . . . that the stat comes from an article that has. . . been . . . thoroughly debunked, by demographers and sociologists and media outlets alike." Susan Faludi, *Backlash: The Undeclared War against American Women* (New York: Crown, 1991), chap. 1, examines the larger anti-feminist context for this message.

3. Garber, "When *Newsweek* 'Struck Terror.'"

4. In similar fashion, women in the late nineteenth century were criticized for educating themselves into sterility and seeking to usurp men's roles. See Margaret Marsh and Wanda Ronner, *The Empty Cradle: Infertility in America from Colonial Times to the Present* (Baltimore: Johns Hopkins University Press, 1996), 75–88.

5. "Fertility and Birth Rates," Child Trends, accessed June 15, 2017, https://www.child trends.org/indicators/fertility-and-birth-rates/; "Live Births, Birth Rates, and Fertility Rates, by Race: United States, 1909–2003," Centers for Disease Control and Prevention, accessed June 15, 2017, https://www.cdc.gov/nchs/data/statab/natfinal2003.annvol1_01.pdf. The baby boom generation, born between 1946 and 1964, included 71.5 million births. The prior generation, born between 1927 and 1945, had just 49.5 million.

6. "First births" among women between the ages of twenty-five and twenty-nine were 31.4 per thousand women in 1975. Women aged thirty to thirty-four had a first-birth rate of 8.0 in 1975, 12.8 in 1980, and 21.2 in 1990.

7. Drew DeSilver, "For Most Workers, Real Wages Have Barely Budged for Decades," Pew Research Center, August 8, 2018, http://www.pewresearch.org/fact-tank/2014/10/09/for-most -workers-real-wages-have-barely-budged-for-decades/.

8. Faludi, *Backlash*, xvii–xix, 71–72, 312–32.

9. Faludi, *Backlash*, 60–61. Quotation on 61.

10. The not-so-subtle message was that feminism, expressed as a drive for professional success, could ruin a woman's chances for love, marriage, and a family. At the time, the premise in the media was that couples were heterosexual; the desire that same-sex couples might have for a family was almost never discussed.

11. Susan Lang, *Women without Children: The Reasons, the Rewards, the Regrets* (New York: Pharos Books, 1991), 43. See also Arthur L. Griel, *Not Yet Pregnant: Infertile Couples in Contemporary America* (New Brunswick, NJ: Rutgers University Press, 1991), 33, and Margarete J. Sandelowski, "Failures of Volition: Female Agencfirst aid kity and Infertility in Historical Perspective," *Signs* 15, no. 3 (Spring 1990): 475–99.

12. The study was based on donor insemination (which is important, because artificial insemination had a longer mean conception time than natural fertilization) in approximately two thousand women with sterile husbands and showed a decline in fertility for women over the age of thirty rather than thirty-five. The decline was slight after thirty and became "marked" after thirty-five. Federation CECOS, D. Schwartz, and M. J. Mayaux, "Female Fecundity as a Function of Age: Results of Artificial Insemination in 2193 Nulliparous Women with Azoospermic Husbands," *New England Journal of Medicine* 306, no. 7 (February 18, 1982): 404–6. The principal prior study that had attempted to directly assess the relationship between age and fecundity had shown only that women over thirty-five seeking a first pregnancy had a median conception time of approximately four months, which was two months longer than for women under twenty-five. That study had been conducted in the 1950s by

Allen Guttmacher. Women under twenty-five had a median "conception time" of 2 months to their first pregnancy, women between the ages of thirty-five and forty-four had a median time to conception of 3.8 months.

13. Alan H. DeCherney and Gertrud Berkowitz, "Female Fecundity and Age" [Editorial], *New England Journal of Medicine* 306, no. 7 (February 18, 1982): 424–26.

14. Edith Brickman and John Beckworth, "Letter to the Editor," *New England Journal of Medicine* 307 (August 5, 1983): 373.

15. For African American women, this was a decline from 16.3 percent in 1965. During the same period, white women saw their infertility rate drop from 10.5 percent in 1965. W. D. Mosher and W. F. Pratt, "Fecundity, Infertility, and Reproductive Health in the United States, 1982," *Vital and Health Statistics* 23:14 (Washington, DC, Government Printing Office, 1987), 4.

16. William D. Mosher and William F. Pratt, "Fecundity and Infertility in the United States, 1965–1988," *Advance Data from Vital and Health Statistics of the National Center for Health Statistics* 192 (December 4, 1990), 1, 3.

17. "About Us," RESOLVE website, accessed June 20, 2017, http://www.resolve.org/about/.

18. After our first book, *The Empty Cradle*, was published in 1996, we were invited to give a talk to a group of gynecologists, and when we mentioned that infertility rates had declined in the United States between 1965 and 1988, the doctors refused to believe us. "That can't be true," they said, incredulously. "We are seeing so many more patients for infertility problems than we did in the 1970s."

19. Robert O. Self, *All in the Family: The Realignment of American Democracy since the 1960s* (New York: Hill and Wang, 2012), esp. 378–82; Tanya Melich, *The Republican War against Women: An Insider's Report from Behind the Lines* (New York: Bantam Books, 1996), passim.

20. Melich, *Republican War against Women*, 150, 165, 167, 171.

21. In 1980, 64 percent of men, compared to 55 percent of women, voted for Reagan. And while his "gender gap" narrowed from seventeen to thirteen points in 1984, it was still significant. Hedrick Smith, "Reagan Easily Beats Carter; Republicans Gain in Congress," *New York Times*, November 5, 1980, http://www.nytimes.com/1980/11/05/politics/05REAG.html; Nate Silver, " 'Gender Gap' Near Historic Highs," *New York Times*, October 21, 2012, http://five thirtyeight.blogs.nytimes.com/2012/10/21/gender-gap-near-historic-highs/. In every presidential election since 1984, with the exception of the first Clinton victory in 1992, the gap has ranged from twelve to twenty points. Even though polls showed that more than two-thirds of Americans still believed in a woman's right to an abortion in the 1980s, it was not at all clear that the issue was important to most voters. Reagan and his party, staking out uncompromising positions on abortion and women's rights, saw the president's sweeping victory as evidence of the political insignificance of both the pro-choice and feminist movements. Melich, *Republican War against Women*, 202, 273.

22. Renate D. Klein, ed., *Infertility: Women Speak Out about Their Experiences of Reproductive Medicine* (Cambridge: Pandora Press, 1989), cover quote.

23. Ellen Hopkins, "Tales from the Baby Factory," *New York Times*, March 15, 1992, SM40.

24. This belief also suggests that motherhood was a heterosexual issue. Although the importance of reproductive technologies, both old and new, to same-sex couples was not evident to the larger society for another decade or so, lesbian couples were clearly availing themselves of donor insemination—a much older technology, of course—to conceive at least as early as the 1980s. See Kara Swanson, *Banking on the Body: The Market in Blood, Milk, and Sperm in Modern America* (Cambridge, MA: Harvard University Press, 2014), 227–29.

25. One of the best overall assessments of the Women's Health Movement is Wendy Kline,

Bodies of Knowledge: Sexuality, Reproduction, and Women's Health in the Second Wave (Chicago: University of Chicago Press, 2010).

26. Fertility Clinic Success Rate and Certification Act of 1992, Pub. L. No. 102-493, 106 Stat. 3149 (October 24, 1992), http://www.gpo.gov/fdsys/pkg/STATUTE-106/pdf/STATUTE-106-Pg3146.pdf.

27. *Human Embryo Transfer: Hearing before the Subcommittee on Investigations and Oversight of the Committee on Science and Technology*, 98th Cong., second session (August 8 and 9, 1984). Hereafter Gore Hearing.

28. Ruth Deech and Anna Smajdor, *From IVF to Immortality: Controversy in the Era of Reproductive Technology* (Oxford: Oxford University Press, 2007), esp. 7–11.

29. Gore Hearing, 2, 20.

30. Gore Hearing, 7.

31. Gore Hearing, 103.

32. Gore Hearing, quote on 236.

33. Gore Hearing, 141.

34. Gore Hearing, 107.

35. *Alternative Reproductive Technologies: Implications for Children and Families. Hearing before the Select Committee on Children, Youth, and Families, House of Representatives*, 100th Cong., first session (May 21, 1987). Hereafter Morrison Hearing.

36. Morrison Hearing, 2.

37. Morrison Hearing, 2.

38. Morrison Hearing, 38, 51, 58.

39. Morrison Hearing, 58, 190, 193.

40. Morrison Hearing, 100–101.

41. Morrison Hearing, 190.

42. *Federal Employee Family Building Act of 1987: Hearing before the Subcommittee on Civil Service of the Committee on Post Office and Civil Service*, 100th Cong., first session (July 23, 1987). Hereafter Schroeder Hearing.

43. Schroeder Hearing, 1.

44. "Schroeder, Patricia Scott," History, Art and Archives: United States House of Representatives," accessed March 18, 2017, http://history.house.gov/People/Listing/S/SCHROEDER,-Patricia-Scott-(S000142)/.

45. A list of the legislation she sponsored, with outcomes, is available at "Representative Patricia Schroeder," Congress.gov, accessed July 6, 2017, https://www.congress.gov/member/patricia-Schroeder/S000142?pageSort=dateOfIntroduction:desc&pageSize=250&q={%22sponsorship%22:%22sponsored%22}.

46. Schroeder Hearing, 9.

47. *Surrogacy Arrangements Act of 1987: Hearing before the Subcommittee of Transportation, Tourism, and Hazardous Materials of the Committee on Energy and Commerce*, 100th Cong., first session, 3–5 (October 15, 1988). Hereafter Luken Hearing. See also Josh Getlin, "Surrogate Motherhood for Pay Is Argued," *Los Angeles Times*, October 16, 1987.

48. Luken Hearing, 6.

49. Susan L. Crockin and Howard W. Jones Jr., *Legal Conceptions: The Evolving Law and Policy of Assisted Reproductive Technologies* (Baltimore: Johns Hopkins University Press, 2010), 77.

50. These data are from our own analysis of the Wyden Questionnaires, hereafter referred to as Marsh-Ronner analysis of Wyden data.

51. *Consumer Protection Issues Involving In Vitro Fertilization Clinics: Hearing before the Subcommittee on Regulation and Business Opportunities of the Committee on Small Business,* 100th Cong., second session, 1 (1988). Hereafter Wyden Hearing, 1988.

52. Wyden Hearing, 1988, p. 2.

53. Nancy Tomes, *Remaking the American Patient: How Madison Avenue and Modern Medicine Turned Patients into Consumers* (Chapel Hill: University of North Carolina Press, 2016), 351. The ideas in this extraordinarily enlightening book can explain a great deal about the changes in medical practice writ large in the twentieth and twenty-first centuries.

54. Michael Lemonick and Dick Thompson, "Trying to Fool the Infertile: Is In-Vitro Fertilization Being Oversold?," *Time* 133, no. 11 (March 13, 1989): 53.

55. Wyden Hearing, 1988, pp. 3–4. Janet Cawley, "Fertility Doctor Betrayed Us, Patients Testify," *Chicago Tribune,* February 24, 1992; "Doctor Is Found Guilty in Fertility Case," *New York Times,* March 5, 1992, https://www.nytimes.com/1992/03/05/us/doctor-is-found-guilty -in-fertility-case.html; "Fertility Doctor Gets Five Years," *New York Times,* May 9, 1992, http:// www.nytimes.com/1992/05/09/us/fertility-doctor-gets-five-years.html.

56. *Consumer Protection Issues Involving In Vitro Fertilization Clinics: Hearing before the Subcommittee on Regulation and Business Opportunities of the Committee on Small Business,* 101st Cong., first session, 2 (1989). Hereafter Wyden Hearing, 1989. See also Fertility Clinic Success Rate and Certification Act of 1992.

57. Fertility Clinic Success Rate and Certification Act of 1992.

58. Fertility Clinic Success Rate and Certification Act of 1992. Even before the passage of the Wyden Act, SART had been publishing aggregated data on success rates. With the passage of this law, clinic-specific data on the success rates for IVF and related reproductive technologies would be made available for patients to consult before choosing a doctor.

59. For a comparison to Britain, see Margaret Foster Riley with Richard A. Merrill, "Regulating Reproductive Genetics: A Review of American Bioethics Commissions and Comparison to the British Human Fertilisation and Embryology Authority," *Columbia Science and Technology Law Review* 6 (2005): 1–64.

60. The estimate is ours, derived from the Wyden data and aggregated SART reports.

Chapter 6. A Lot of Money Being Made

1. Arthur Caplan quoted in Susan Kelleher and Kim Christensen, "Baby Born after Doctor Took Eggs without Consent," *Orange County Register,* May 19, 1995, http://www.pulitzer .org/winners/staff-37.

2. The *Orange County Register* won the 1996 Pulitzer Prize for Investigative Reporting for its series of stories on the Asch scandal. The stories can be found on the Pulitzer Prize website at http://www.pulitzer.org/winners/staff-37. The two stories quoted and cited here were published on May 19, 1995, and November 4, 1995.

3. DeCherney said that he was told the figure was closer to $90 million. Alan DeCherney, interview with the authors, November 2, 2015.

4. Hilary Gilson, "Ricardo Hector Asch (1947–)," *Embryo Project Encyclopedia,* June 10, 2009, http://embryo.asu.edu/handle/10776/1936. Asch and Balmaceda were not the first to try the technique, but they were the first to bring about a live birth using it and therefore are credited with its development.

5. On Greenblatt's discovery, see Margaret Marsh and Wanda Ronner, *The Fertility Doctor: John Rock and the Reproductive Revolution* (Baltimore: Johns Hopkins University Press, 2008), 194. AMI stands for American Medical International, the name of center's location in Garden Grove. A second location opened in 1989, and the following year the practice moved into new facilities built in Irvine in the hospital complex. *Consumer Protection Issues Involving In Vitro Fertilization Clinics: Hearing before the Subcommittee on Regulation and Business Opportunities of the Committee on Small Business*, 100th Cong., second session, 109 (1988). Hereafter Wyden Hearing, 1988. See also Hilary Gilson, "Center for Reproductive Health (1986–1995)," *The Embryo Project Encyclopedia*, September 30, 2009, http://embryo.asu.edu/handle/10776/1946.

6. Mary Dodge and Gilbert Geis, *Stealing Dreams: A Fertility Clinic Scandal* (Boston: Northeastern University Press, 2003), 30–31.

7. Dodge and Geis, *Stealing Dreams*, 30–31. Asch's income is impossible to estimate, but UC Irvine claimed that he and his partners failed to report more than $7 million in revenue. "Fertility Clinic Doctors Owe Campus $2.47 Million, Audit Says," *Los Angeles Times*, March 26, 1998.

8. Geoffrey Cowley and Andrew Murr, "Ethics and Embryos," *Newsweek* 125, no. 24 (June 12, 1995): 66; Dodge and Gies, *Stealing Dreams*, 116.

9. As of 2016, Stone had a private practice in Calexico, California. Dodge and Geis, *Stealing Dreams*, 156.

10. Dodge and Geis, *Stealing Dreams*, 147, 158.

11. The NIH had sent an investigative team because there were possible research misconduct implications to the allegations. DeCherney, interview with the authors.

12. DeCherney, interview with the authors.

13. DeCherney, interview with the authors.

14. Dodge and Geis, *Stealing Dreams*, 29–31.

15. By 2005, fewer than 1 percent of cycles in American fertility clinics used GIFT.

16. DeCherney, interview with the authors.

17. Kathryn Wexler, "Missing Eggs Leave Empty Arms," *Washington Post*, December 31, 1995, https://www.washingtonpost.com/archive/politics/1995/12/31/missing-eggs-leave-empty-arms/7af7779f-3d72-4f75-8d81-5e2fddfaaa4e/?utm_term=.e2d596e3f708.

18. Dodge and Geis, *Stealing Dreams*, 16–17.

19. Michelle Nicolosi, "Corona Parents Make Ordeal Public," *Orange County Register*, June 8, 1995, http://www.pulitzer.org/winners/staff-37.

20. Dodge and Geis, *Stealing Dreams*, 16–17; DeCherney, interview with the authors; Cowley and Murr, "Ethics and Embryos," 66.

21. Julie Marquis, "Fertility Doctor Denies Role in Errors," *Los Angeles Times*, January 20, 1996, http://articles.latimes.com/1996-01-20/news/mn-26676_1_fertility-doctor.

22. "Fugitive in UC Irvine Fertility Scandal Arrested in Mexico City; U.S. Hopes to Extradite Him," *Los Angeles Times*, December 27, 2010, http://latimesblogs.latimes.com/lanow/2010/12/uci-fertility-scandal-ricardo-asch-arrest-mexico-city-extradition.html.

23. Kim Christensen, "Doctor with Ties to Fertility Scandal Won't Be Extradited by Mexico," *Los Angeles Times*, April 1, 2011, http://articles.latimes.com/2011/apr/01/local/la-me-0401-asch-20110401.

24. Dodge and Geis, *Stealing Dreams*, 138–39. Quotation is from authors paraphrasing Balmareda.

25. The most prominent of these sanctioned IVF clinics was a relatively new for-profit chain called IVF-Australia, whose main operation was in Port-Chester, New York. See "Complaint

in the Matter of NME Hospitals, Inc.," accessed October 4, 2018, https://www.ftc.gov/sites/default/files/documents/commission_decision_volumes/volume-113/volume113_1115-1182.pdf, 1171–82.

26. *Consumer Protection Issues Involving In Vitro Fertilization Clinics: Hearing before the Subcommittee on Regulation and Business Opportunities of the Committee on Small Business*, 101st Congress, first session, 136 (1989). Hereafter Wyden Hearing, 1989.

27. Leslie Laurence, "The Bargain Baby-Maker," *New York Magazine*, accessed September 8, 2017, http://nymag.com/nymetro/health/features/2163/.

28. "Interview: Dr. Mark Sauer," *Frontline: Making Babies* (1999), accessed September 8, 2017, http://www.pbs.org/wgbh/pages/frontline/shows/fertility/interviews/sauer.html.

29. From authors' analysis of SART data for 2000.

30. Out of a total of about 195 clinics, 163 clinics reported their data with the Society for Reproductive Technology. We are using only the IVF data, not data for GIFT or ZIFT (zygote intrafallopian transfer, a variant of GIFT in which the egg was fertilized and then the fertilized egg, called a zygote, was placed in the fallopian tube). A total of 2,104 women, or 15.5 percent, delivered one or more babies. Just 2 percent of IVF patients became pregnant with donor eggs. Success rates are higher when donor eggs are used, and donor egg births were 3.8 percent of the total. The number of women is approximate; Medical Research International, Society for Assisted Reproductive Technology, and American Fertility Society, "In Vitro Fertilization-Embryo Transfer (IVF-ET) in the United States: 1989 Results from the IVF-ET Registry," *Fertility and Sterility* 55, no. 1 (January 1991): 14–23. Hereafter 1989 Results.

31. There are no exact data on the numbers of babies born after IVF in the United States between the birth of Elizabeth Jordan Carr in 1981 and the end of the decade. SART did not begin collecting such data until 1985. Between then and 1989, the data reported 4,752 births resulting from IVF. Before that, we can only estimate, and from our larger reading, we estimate perhaps 200 or so, giving us roughly 5,000 births over the course of the 1980s. We calculated the figure for the 1990s by adding up the deliveries in each SART report for the decade. The exact number for the 1990s was 80,921.

32. Exactly 21,904 women gave birth in 1999. Success rates, in terms of live births per egg retrieval, were 29.4 percent. In total, 13.8 percent of births involved the use of donor eggs, and 1.1 percent of the births were to gestational surrogates. Because of changes in the way that data were gathered beginning in the mid-1990s, it is not possible precisely to compare birth outcomes between 1989 and 1999. Society for Assisted Reproductive Technology and the American Society for Reproductive Medicine, "Assisted Reproductive Technology in the United States: 1999 Results Generated from the American Society for Reproductive Medicine / Society for Assisted Reproductive Technology Registry," *Fertility and Sterility* 78, no. 5 (November 2002): 918–31. Hereafter 1999 Results. See also 1989 Results, 14–23. The number of women is approximate. By 1999, IVF had become the dominant technology. GIFT and ZIFT had declined to just 2 percent of all deliveries by 1999, down from 25 percent in 1989. See 1999 Results and 1989 Results.

33. The mean age of first birth in women was rising. In 1970, it was 21.4. By 2000, it was 24.5. T. J. Mathews, MS, and Brady E. Hamilton, PhD, "Mean Age of Mother, 1970–2000," *National Vital Statistics Report* 51, no. 1 (December 11, 2002): 10, https://www.cdc.gov/nchs/data/nvsr/nvsr51/nvsr51_01.pdf. See also Joyce A. Martin et al., "Births: Final Data for 2000," *National Vital Statistics Report* 50, no. 5 (February 12, 2002): 1–3; Amara Bachu and Martin O'Connell, *Fertility of American Women: June 2000* (Washington, DC: US Census Bureau, 2001), 2, https://www.census.gov/prod/2001pubs/p20-543rv.pdf.

34. Sherrye Henry, "What It Takes to Bear a Child at Forty and Older," *Washington Post*, January 19, 1992, N16. The headline notwithstanding, this was an article about mothers aged thirty-five and older.

35. J. Abma, A. Chandra, W. Mosher, L. Peterson, and L. Piccinino, "Fertility, Family Planning, and Women's Health: New Data from the 1995 National Survey of Family Growth," *National Center for Health Statistics: Vital Health Statistics* 23, no. 19 (1997): 6–7.

36. In all, 9 percent received advice, 6 percent had medical tests, 4 percent took ovulation induction drugs, 2 percent had surgery or another treatment for fallopian tube disease, and 1.4 percent underwent artificial insemination. Anjani Chandra, Casey E. Copen, and Elizabeth Hervey Stephen, "Infertility Service Use in the United States: Data from the National Survey of Family Growth, 1982–2010," *National Health Statistics Report* 73 (January 22, 2014): 15.

37. Abma et al., "Fertility, Family Planning, and Women's Health," 65.

38. Lisa Hope Harris, "Challenging Conceptions: A Clinical and Cultural History of In Vitro Fertilization in the United States" (PhD diss., University of Michigan, 2006). See p. 332 for the 1990s study. Dorothy Roberts, *Killing the Black Body: Race, Reproduction, and the Meaning of Liberty* (New York: Vintage, 1999), 253. [Originally published 1997.]

39. Cindy Loose, "A Holiday Comes to Life: Mom's Celebration Was 4 Years in Making," *Washington Post*, May 9, 1993, A1.

40. Dan Beyers, " 'Miracle Babies' Reunite at Md. In Vitro Center," *Washington Post*, October 22, 1990, D3.

41. Susan L. Crockin and Howard W. Jones Jr., *Legal Conceptions: The Evolving Law and Policy of Assisted Reproductive Technologies* (Baltimore: Johns Hopkins University Press, 2010), 77.

42. Richard Marrs, interview with authors, June 23, 2017; testimony by Gary Hodgen, *Alternative Reproductive Technologies: Implications for Children and Families*. Hearing before the Select Committee on Children, Youth, and Families, 100th Congress, first session (May 21, 1987).

43. Marrs, interview with authors. Regarding Penn, Steven Sondheimer, interview with authors, February 29, 2016. According to Sondheimer, Penn still recruits its own donors and does not use agencies.

44. American Society for Reproductive Medicine and the Society for Assisted Reproductive Technology, "Assisted Reproductive Technology in the United States and Canada: 1994 Results Generated from the American Society for Reproductive Medicine / Society for Assisted Reproductive Technology Registry," *Fertility and Sterility* 66, no. 5 (November 1996): 701–2; American Society for Reproductive Medicine and the Society for Assisted Reproductive Technology, "Assisted Reproductive Technology in the United States and Canada: 1993 Results Generated from the American Society for Reproductive Medicine / Society for Assisted Reproductive Technology Registry," *Fertility and Sterility* 64, no.1 (July 1995): 17–18.

45. Rene Almeling, *Sex Cells: The Medical Market for Eggs and Sperm* (Berkeley: University of California Press, 2011), 42–43.

46. Kathleen Doheny, "A Priceless Possibility," *Los Angeles Times*, March 19, 1995, http://articles.latimes.com/1995-03-19/news/ls-44521_1_mark-sauer. At first, like most of his colleagues, Sauer believed that women over the age of forty would have difficulty conceiving even with donor eggs, but then he did a study of thirty-one patients who used egg donors. The patients ranged from age twenty-four to forty-four and just a few were older than forty, but he and his colleagues found a high rate of pregnancy even in the older patients. Mark Sauer et al.,

"Establishment of a Non-Anonymous Donor Oocyte Program: Preliminary Experience at the University of Southern California," *Fertility and Sterility* 52, no. 3 (September 1989): 433–36; Mark Sauer et al., "Oocyte and Pre-Embryo Donation to Women with Ovarian Failure: An Extended Clinical Trial," *Fertility and Sterility* 55, no. 1 (January 1991): 39–43. Sauer's research at the University of Southern California was approved by its institutional review board. Mark V. Sauer, Richard J. Pauslon, and Rogerio A. Lobo, "A Preliminary Report on Oocyte Donation Extending Reproductive Potential to Women over 40," *New England Journal of Medicine* 323, no. 17 (1990): 1157–60; M. V. Sauer, R. J. Paulson, and R. A. Lobo, "Pregnancy after Age 50: Application of Oocyte Donation to Women after Natural Menopause," *The Lancet* 341, no. 8841 (1993): 321–23, https://www.ncbi.nlm.nih.gov/pubmed/8094110; Richard Paulson and Mark Sauer, "Regulation of Oocyte Donation to Women over the Age of 50: A Question of Reproductive Choice," *Journal of Assisted Reproduction and Genetics* 11, no. 4 (April 1994): 177–82.

47. Doheny, "Priceless Possibility."

48. Shari Roan, "Woman Gives Birth at 63: Ethical Questions Raised," *Los Angeles Times*, April 24, 1997, http://articles.latimes.com/1997-04-24/news/mn-51898_1_ethical-question.

49. Society for Assisted Reproductive Technology and the American Society for Reproductive Medicine, "Assisted Reproductive Technology in the United States and Canada: 1994 Results Generated from the American Society for Reproductive Medicine / Society for Assisted Reproductive Technology Registry," *Fertility and Sterility* 66, no. 5 (November 1996): 701–2; Society for Assisted Reproductive Technology and the American Society for Reproductive Medicine, "Assisted Reproductive Technology in the United States and Canada: 1993 Results Generated from the American Society for Reproductive Medicine / Society for Assisted Reproductive Technology Registry," *Fertility and Sterility* 64, no. 1 (July 1995): 17–18.

50. Dorene Weinstein, "Whatever Happened To: Surrogate Grandmother," *Argus Leader*, October 7, 2014, http://argusne.ws/1rIA1Nm; Gina Kolata, "When Grandmother Is Mother, until Birth," *New York Times*, August 4, 1991, http://www.nytimes.com/1991/08/05/us/when-grandmother-is-the-mother-until-birth.html.

51. She became the first grandmother in the United States and just the second in the world to bear a daughter's children. Weinstein, "Whatever Happened"; Kolata, "When Grandmother Is Mother." On the first grandmother to serve as a gestational surrogate, in South Africa, see John D. Battersby, "Woman Is Carrying Her Daughter's Babies," *New York Times*, April 9, 1987, A1.

52. Kolata, "When Grandmother Is Mother."

53. Weinstein, "Whatever Happened."

54. An authoritative look at the two types of surrogacy is Susan Markens, *Surrogate Motherhood and the Politics of Reproduction* (Berkeley: University of California Press, 2007).

55. Sorkow ruled on March 31, 1987, that the surrogacy contract was valid, awarded custody to the Sterns, and terminated Whitehead's parental rights. Robert Hanley, "Father of Baby M Granted Custody," *New York Times*, April 1, 1987, http://www.nytimes.com/1987/04/01/nyregion/father-of-baby-m-granted-custody-contract-upheld-surrogacy-is-legal.html?pagewanted=1.

56. "Excerpts from Decision by New Jersey Supreme Court in the Baby M Case," *New York Times*, February 4, 1988, http://www.nytimes.com/1988/02/04/nyregion/excerpts-from-decision-by-new-jersey-supreme-court-in-the-baby-m-case.html?pagewanted=2.

57. Hanley, "Father of Baby M Granted Custody."

58. Bonnie Johnson, "And Baby Makes Four: For the First Time a Surrogate Bears a Child Genetically Not Her Own," *People* 27, no. 18 (May 4, 1987): http://www.people.com/people /archive/article/0,,20096199,00.html.

59. Johnson, "And Baby Makes Four."

60. James M. Goldfarb, Cynthia Austin, Barry Peskin, Hannah Lisbona, Nina Desai, and J. Ricardo Loret de Mola, "Fifteen Years Experience with an In Vitro Fertilization Surrogate Gestational Pregnancy Programme," *Human Reproduction* 15, no. 5 (2000): 1075–78.

61. Fifteen of Utian patients suffered from the same medical condition as Arlette Schweitzer's daughter, where the uterus was congenitally absent. More than half the women seeking a surrogate (fifty-eight) were between the ages of thirty-one and thirty-five. Fifteen were in their twenties, and ten were between forty and forty-five. Goldfarb et al., "Fifteen Years Experience." "Suitability" statement is on p. 1076.

62. Markens, *Surrogate Motherhood*, 173–76.

63. Tina Nguyen, "Fertility Clinic Plaintiffs Are Still Grateful," *Los Angeles Times*, November 11, 1995, http://articles.latimes.com/1995-11-11/local/me-1864_1_fertility-clinic.

64. Markens, *Surrogate Motherhood*, 132–35.

65. Johnson, "And Baby Makes Four."

66. Aileen Ballantyne, "Bourn Hall Will Offer Babies without Birth," *Sunday Times,* August 19, 1990.

67. Markens, *Surrogate Motherhood*, 115–18.

68. Almeling, *Sex Cells*, 150.

69. According to Wulf Utian, "Surrogate gestational carriers are significantly different" from traditional surrogates, "since the surrogate has no genetic link to the fetus." Goldfarb et al., "Fifteen Years Experience," 1075.

70. Of those who intended to use these embryos later, 94 percent had done so by the time of the study. Catherine V. Hounshell and Ryszard J. Chetkowski, "Donation of Frozen Embryos after In Vitro Fertilization Is Uncommon," *Fertility and Sterility* 66, no. 5 (November 1996): 837–38. The patients froze their embryos between 1989 and 1991.

71. Bette Sehnert and Ryszard J. Chetkowski, "Secondary Donation of Frozen Embryos Is More Common after Pregnancy Initiation with Donated Eggs Than after In Vitro Fertilization-Embryo Transfer and Gamete Intrafallopian Transfer," *Fertility and Sterility* 69, no. 2 (February 1998): 350–52. Fifty-two of them retained their embryos for later. Of the other sixteen, eleven donated their "spare" embryos and five discarded them. See 1999 Results.

72. Gina Kolata, "Clinics Selling Embryos Made for 'Adoption,'" *New York Times*, November 23, 1997, 1, 34.

73. Kolata, "Clinics Selling Embryos," 34.

74. In 1909, a physician claimed that twenty-five years earlier he had witnessed prominent Philadelphia surgeon and Jefferson Medical School professor James Pancoast use the sperm of one of his medical students to inseminate a patient whose husband was azoospermic. By then, Pancoast was dead and no other witnesses came forth, so we have no way of knowing whether this story was true. Apocryphal or true, the story was widely believed. Edward Bliss Foote was an 1858 graduate of the medical school of the University of Pennsylvania and New York practitioner who dispensed medical advice by mail and in two popular books. In one, published in 1870, he advised readers on the use of artificial insemination to treat infertility. In many cases, he said, the insemination could be performed with the husband's sperm. But if the semen had no spermatozoa, his said, "the male germs must be obtained" from another man. Foote sold

his "impregnating syringe," with complete instructions, for $5.00. Marsh and Ronner, *Empty Cradle*, 67–69, 94–95.

75. Marsh and Ronner, *Empty Cradle*, 166–67.

76. John Rock had been using donor insemination since at least the 1940s, and possibly earlier. Marsh and Ronner, *Empty Cradle*, 198–206. On the modern sperm bank, see Kara W. Swanson, *Banking on the Body: The Market in Blood, Milk, and Sperm in Modern America* (Cambridge, MA: Harvard University Press, 2014).

77. "How Common Is Male Infertility, and What Are Its Causes?," Eunice Kennedy Shriver National Institute of Child Health and Human Development, accessed October 4, 2018, https://www.nichd.nih.gov/health/topics/menshealth/conditioninfo/infertility.

78. P. Devroey and A. Van Steirteghem, "A Review of Ten Years Experience of ICSI," *Human Reproduction Update* 10 (2004): 19. The first procedure was called PZD, for partial zona dissection, and the second was called SUZI, for subzonal insemination.

79. Offer Harari et al., "Intracytoplasmic Sperm Injection: A Major Advance in the Management of Severe Male Subfertility," *Fertility and Sterility* 64, no. 2 (1995): 360–68. For the report from Van Steirteghem's group, see G. Palermo, H. Joris, P Devroey, and A. C. Van Steirteghem, "Pregnancies after Intracytoplasmic Injection of Single Spermatozoon into an Oocyte," *Lancet* 340, no. 8810 (1992): 17–18.

80. SART, "Assisted Reproductive Technology in the United States and Canada: 1995 Results Generated from the American Society for Reproductive Medicine / Society for Assisted Reproductive Technology Registry," *Fertility and Sterility* 69, no. 3 (March 1998): 395. See also 1999 Results. Not all women who received ICSI had infertile partners. Some clinics used the technique for other reasons. As the report said, "When all IVF cycles are classified as 'male factor infertility' and 'other diagnosis,' patients with male factor infertility had the same delivery rate per retrieval (29.4%) as those with other diagnoses (29.4%)."

81. To ensure that sperm was free of the HIV virus, it needed to be frozen and quarantined, then tested after an appropriate period of time had passed. Sperm banks had been in existence since the 1970s, but until the AIDS crisis, physicians had preferred to use fresh sperm from donors they themselves recruited. From the late 1980s on, because of HIV, this proved impossible, and sperm banks have been dominant ever since. Lesbian couples who wanted to have children in the 1970s and 1980s would often turn to male friends or the few gay-friendly medical clinics. The development of easily accessible sperm banks gave them other choices by the 1990s. Bridget Gurtler, "Synthetic Conception: Artificial Insemination and the Transformation of Reproduction and Family in Nineteenth and Twentieth Century America" (PhD diss., Rutgers University, 2013), chap. 5.

82. Almeling, *Sex Cells*, 32–33. Swanson, *Banking on the Body*.

83. Serena H. Chen, Claudia Pascale, Maria Jackson, Mary Ann Szvetecz, and Jacques Cohen, "A Limited Survey-Based Uncontrolled Follow-Up Study of Children Born after Ooplasmic Transplantation in a Single Centre," *Reproductive BioMedicine Online* 33 (October 2016): 738.

84. Kim Tingley, "The Brave New World of Three-Parent I.V.F.," *New York Times Magazine*, June 27, 2014, https://www.nytimes.com/2014/06/29/magazine/the-brave-new-world-of-three-parent-ivf.html.

85. Chen et al., "Limited Survey-Based Uncontrolled Follow-Up Study," 737–44. Of the thirteen women who succeeded in giving birth, the youngest was 30.5 and the oldest was 39.5, with an average age of 35.6.

86. With one exception, the couples were unwilling to use donor eggs for their pregnancies. The couple who constituted the one exception had twins. The wife had both a donor's egg and one with cytoplasmic transfer implanted at the same time. Both "took," and one of her twins was from the donor egg, the other from her own egg with a donor's cytoplasm. Chen et al., "Limited Survey-Based Uncontrolled Follow-Up Study," 738.

87. Alison Freehling, "Making a Good Egg," *Daily Press*, August 24, 1998.

88. Philip Sherwell, "Three-Parent Babies," *The Telegraph*, August 30, 2014, http://www .telegraph.co.uk/news/health/11065448/Three-parent-babies-I-thought-I-would-never-have -my-own-baby.-I-was-distraught.html#.

89. Jason A. Barritt, Carol A. Brenner, Henry E. Malter, and Jacques Cohen, "Mitochondria in Human Offspring Derived from Ooplasmic Transplantation: Brief Communication," *Human Reproduction* 16, no. 3 (March 1, 2001): 513–16, https://doi.org/10.1093/humrep/16.3.513.

90. Faye Flamm, "Cloning Technique Helps Woman Conceive," *Philadelphia Inquirer*, July 22, 1997; Tingley, "Brave New World," 28.

91. Chen et al., "Limited Survey-Based Uncontrolled Follow-Up Study," 738.

92. National Institutes of Health, *Report of the Human Embryo Research Panel*, vol. 1 (Bethesda, MD: National Institutes of Health, 1994), v.

93. Ronald M. Green, "The Human Embryo Research Panel: Lessons for Public Ethics," *Cambridge Quarterly of Healthcare Ethics* 4, no. 4 (1995): 502–15.

94. Green, "Human Embryo Research Panel," 502.

95. "Dickey-Wicker Amendment, 1996," *Embryo Project Encyclopedia*, August 27, 2010, https://embryo.asu.edu/pages/dickey-wicker-amendment-1996.

96. Andrea L. Kalfoglou, "Looking Back, Looking Forward: The Legacy of the National Advisory Board on Ethics in Reproduction (NABER), *Women's Health Issues* 10, no. 3 (May–June 2000): 92–104. Quote on p. 99.

97. Exec. Order No. 12,975, 60 Fed. Reg. 193 (October 3, 1995), 52,063–65, https://bioethics archive.georgetown.edu/nbac/about/eo12975.htm.

98. Victoria C. Wright, Laura A. Schieve, Meredith A. Reynolds, and Gary Jeng, "Assisted Reproductive Technology Surveillance—United States, 2000," *Surveillance Summaries* 52, no. SS09 (August 29, 2003): 1–16, http://www.cdc.gov/mmwr/preview/mmwrhtml/ss5520 9a1.htm; "Live Births and Birth Rates, by Year," Infoplease.com, accessed January 6, 2016, http://www.infoplease.com/ipa/A0005067.html. The exact number was 4,058,814. Emma Innes, "Five Million Babies Have Now Been Born by IVF and HALF since 2007, New Figures Confirm," *Daily Mail*, October 16, 2013, http://www.dailymail.co.uk/health/article-2462640 /Five-million-babies-born-IVF--HALF-2007.html.

Chapter 7. Beyond Infertility

1. Tatiana Boncompagni, "Are You as Fertile as You Look?," *New York Times*, September 1, 2011.

2. Liza Mundy, *Everything Conceivable: How Assisted Reproduction Is Changing Our World* (New York: Anchor Books, 2008), 24–27, 57–60.

3. Cynthia Hanson, "We Can't Get Pregnant and It's Driving Us Apart," *Ladies' Home Journal* (March 2008): http://www.more.com/love-sex/marriage/can-marriage-be-saved/we -cant-get-pregnant-and-its-driving-us-apart.

4. Cynthia Hanson, "My Infertility Is Ruining Our Marriage," *Ladies' Home Journal* (Sep-

tember 2011): http://www.more.com/love-sex/marriage/can-marriage-be-saved/my-infertility
-ruining-our-marriage.

5. Hanson, "My Infertility Is Ruining Our Marriage."

6. Hanson, "My Infertility Is Ruining Our Marriage"; Mundy, *Everything Conceivable*,
57–60; Hanson, "We Can't Get Pregnant."

7. They were part of the magazine's popular series called "Can This Marriage Be Saved?"
The series had run for decades. These two stories appeared in 2008 and 2011.

8. Kristin Celello, *Making Marriage Work: A History of Marriage and Divorce in the
Twentieth-Century United States* (Chapel Hill: University of North Carolina Press, 2009),
99–104.

9. Mundy, *Everything Conceivable*, back cover of the paperback edition.

10. The 2015 data are the most recent available, as of this writing, from the Centers for Disease Control and Prevention (CDC).

11. Data calculated from individual clinic numbers. See "ART Success Rates," Centers for
Disease Control, last updated May 16, 2018, https://www.cdc.gov/art/artdata/index.html.

12. Alan DeCherney, interview with the authors, November 2, 2015. Here is his full quote
on "passing the torch": "I wrote the review of your book [*The Fertility Doctor*] in the *New
England Journal of Medicine*. While I was preparing the review, Luigi [Mastroianni] was dying. I spent an afternoon with him several days before his death. We talked about the Rock
book, and it was a wonderful experience for me to be with him. It was like Rock passed the
torch to him, and he passed the torch to me. It was very moving."

13. From the authors' interviews with Richard Marrs, June 23, 2017; Martin Quigley, June 9,
2017; Christos Coutifaris, December 16, 2015; Edward Wallach, April 6, 2016; and Esther
Eisenberg, March 4, 2017. Eisenberg was on the faculty at Vanderbilt when its programs were
relocated to private practices.

14. Marrs, interview with the authors.

15. Data calculated from individual clinic numbers; see "ART Success Rates, 2015."

16. We have not been able to discover the typical earnings of directors of for-profit fertility
centers. Universities, however, as non-profit organizations, must report the compensation of
their top earners. Cornell reports that, in addition to the director, Zev Rosenwaks, two other
physicians in the practice, Hey-Joo Kang and Steven Spandorfor, had total compensation (including both salary and benefits) of $3.9 million and $3.5 million, respectively. See Drew
Musto and Nicholas Bogel-Burrough, "No Black or Hispanic Employees among Highest Paid
on Cornell's IRS Return," *Cornell Sun*, September 1, 2017, http://cornellsun.com/2017/09/07
/highest-paid-on-cornells-irs-return-mostly-white-men/.

17. Data from April 2017; the exact number is $333,824. See Andrea Clement Santiago, "Average Doctors' Salaries by Specialty," Verywell Health, last updated February 24, 2018, https://
www.verywell.com/doctors-salaries-by-specialty-1736005.

18. Debora Spar, *The Baby Business: How Money, Science, and Politics Drive the Commerce of
Conception* (Boston: Harvard Business School Press, 2006), xv.

19. Harris-Williams & Co., *Fertility Market Overview* (Boston: Harris-Williams, May 2015):
http://www.harriswilliams.com/sites/default/files/content/fertility_industry_overview
_-_2015.05.19_v10.pdf; Ed Sealover, "Colorado Reproductive Clinic Gets Major Cash Infusion," *Denver Business Journal*, August 4, 2015, https://www.bizjournals.com/denver/news/2015
/08/04/colorado-reproductive-clinic-gets-major-cash.html. For a list of centers, see the CCRM
website, accessed January 2, 2018, https://www.ccrmivf.com/#.

20. These figures come from the National Survey of Family Growth. The 2010 survey was

taken between 2006 and 2010. It conducts periodic surveys on multiple aspects of fertility and infertility. See Anjani Chandra, Casey E. Copen, and Elizabeth Hervey Stephen, "Infertility and Impaired Fecundity in the United States, 1982–2010: Data from Chandra et al., "Infertility and Impaired Fecundity in the United States, 1982–2010," *National Health Statistics Report* 67 (August 14, 2013): 6, https://www.cdc.gov/nchs/data/nhsr/nhsr067.pdf. Impaired fecundity was on the rise until 2002, from 11 percent in 1982 to 15 percent in 2002, then down to 12 percent in 2010. According to this report, that rise was driven primarily by an increase in the numbers of women who said it was physically difficult or dangerous for them to have a baby, not by those who could not conceive.

21. In the 1990s, women were treated if they had both CMT (cervical motion tenderness) and uterine adrenal tenderness, but the new criteria said that treatment should occur if they had just one of them. Monika Goyal, "No Increase in ER Diagnosis in Adolescents 14–21," *Journal of Adolescent Health* 53, no. 2 (August 2013): 240–52; Chandra et al., "Infertility and Impaired Fecundity."

22. Anjani Chandra, C. E. Copen, and E. H. Stephen, "Infertility Service Use in the United States: Data from the National Survey of Family Growth, 1982–2010," *National Health Statistics Report* 73 (January 22, 2014): 6–7. Medical help included advice (9 percent), medical tests for the woman and/or her spouse (7 percent), ovulation induction medication (6 percent), artificial insemination (1.7 percent).

23. See "Who Has Infertility?," Resolve.org, accessed November 14, 2017, http://resolve.org/infertility-101/what-is-infertility/fast-facts/, and "Quick Facts about Infertility," ReproductiveFacts.org, accessed January 25, 2018, http://www.reproductivefacts.org/faqs/quick-facts-about-infertility/.

24. The exact numbers were 21,904 and 60,778, respectively. In 1999, a tiny percentage of cycles used GIFT or ZIFT, a technology similar to GIFT. In ZIFT, the eggs are fertilized and the resulting embryo (or zygote) is transferred to the fallopian tubes. Society for Assisted Reproductive Technology and the American Society for Reproductive Medicine, "Assisted Reproductive Technology in the United States: 1999 Results Generated from the American Society for Reproductive Medicine/Society for Assisted Reproductive Technology Registry," *Fertility and Sterility* 78, no. 5 (November 2002): 918. Hereafter 1999 Results. Of the 370 clinics, 360 reported their data to the Society for Assisted Reproductive Technology; information is based on that reporting group. For 2015, the number included 45,779 cycles where the intent was to freeze all the resultant eggs or embryos, not to implant them. The remaining 186,157 cycles were done with the intent to transfer at least one egg or embryo. For the take-home baby rate, we are counting only cycles for which an embryo transfer was intended. Centers for Disease Control and Prevention, American Society for Reproductive Medicine, Society for Assisted Reproductive Technology, *2015 Assisted Reproductive Technology National Summary Report* (Atlanta, GA: US Department of Health and Human Services, 2017), 3, 7. For 2015 statistics, see https://www.cdc.gov/art/pdf/2015-report/ART-2015-National-Summary-Report.pdf#page=11.

25. Because women undergo IVF procedures, clinics record only the ages of these women, not their male partners.

26. See 1999 Results; percentage calculated from figures on p. 919. For 2015, the percentage was calculated from *2015 Assisted Reproductive Technology National Summary*, 52.

27. James P. Toner et al., "Society for Assisted Reproductive Technology and Assisted Reproductive Technology in the United States: A 2016 Update," *Fertility and Sterility* 106, no. 3 (September 2016): 541–46. Table on p. 544.

28. In 2000, 43 percent of cycles involved women under thirty-five, 43 percent were done in women between thirty-five and thirty-nine, and 13 percent of cycles involved women forty and over. In 2015, women under thirty-five underwent 38 percent of the cycles, women between thirty-five and forty 40 percent, and women forty and over 22 percent. See figure A.1; percentages are rounded.

29. These data are from 2015 and can be viewed at "ART Success Rates," Centers for Disease Control and Prevention, accessed January 2, 2018, https://www.cdc.gov/art/artdata/index.html.

30. Women with the least education had children at younger ages. About 88 percent of women between twenty-five and thirty-four, who had not completed high school, had children in 2010. Sabrina Tavernese, "Census Data Reveals a Shift in Patterns of Childbearing," *New York Times*, May 10, 2011; Gladys Martinez, Kimberly Daniels, and Anjani Chandra, "Fertility of Men and Women Aged 15–44 Years in the United States: National Survey of Family Growth, 2006–2010," *National Health Statistics Report* 51 (April 12, 2012): 13, https://www.cdc.gov/nchs/data/nhsr/nhsr051.pdf#page=1&zoom=auto,-13,792.

31. In 2015, there were 11 births overall per one thousand women between the ages of forty and forty-four, the highest rate since 1966, when women of that age were most likely to be completing their families rather than starting them. Women older than forty-five had a birth rate of 0.7 births per one thousand women, including 754 births to women over age fifty. Joyce A. Martin, Brady E. Hamilton, Michelle J. K. Osterman, Anne K. Driscoll, and T. J. Mathews, "Births: Final Data for 2015," *National Vital Statistics Reports* 66, no. 1 (January 5, 2017): 2–5, https://www.cdc.gov/nchs/data/nvsr/nvsr66/nvsr66_01.pdf.

32. Claire Cain Miller, "Children Hurt Women's Earnings, but Not Men's," *New York Times*, February 5, 2018, https://www.nytimes.com/2018/02/05/upshot/even-in-family-friendly-scandinavia-mothers-are-paid-less.html.

33. In 2012, 23 percent of men and 17 percent of women over the age of twenty-five were unmarried. In 1960, the median age of first marriage was twenty-one for women and twenty-four for men. Twenty years later, the median age had gone up to just over twenty-two for women and twenty-five for men. And by the first year of the new millennium, the median age at first marriage for women was just under 25 and 26.5 for men. Wendy Wang and Kim Parker, *Record Share of Americans Have Never Married as Values, Economics and Gender Patterns Change* (Washington, DC: Pew Research Center's Social and Demographic Trends Project, 2014), 4–5.

34. Diana B. Elliott, Kristy Krivickas, Matthew W. Brault, and Rose M. Kreider, *Historical Marriage Trends from 1890–2010: A Focus on Race Differences*, Working Paper SEHSD-WP 2012-12 (Washington, DC: US Census Bureau, May 2012), figs. 1–3, https://www.census.gov/library/working-papers/2012/demo/SEHSD-WP2012-12.html. In 2010, 86 percent of men and 88 percent of women over the age of thirty-five had married, and the other 4 percent married between thirty-six and forty-four.

35. Stephanie Coontz, "Is Marriage Becoming Obsolete?," *CNN*, November 22, 2010, http://www.cnn.com/2010/OPINION/11/22/coontz.marriage.pew/index.html.

36. American College of Obstetricians and Gynecologists Committee on Gynecologic Practice and the Practice Committee of the American Society for Reproductive Medicine, "Female Age-Related Fertility Decline: Committee Opinion No. 589," *Fertility and Sterility* 101, no. 3 (March 2014): 633–34.

37. Jon Saraceno, "Last Chance Babies: Sure They Do It, But Should They?," *AARP Bulletin* (January–February 2015): 22, 24.

38. Boncompagni, "Are You as Fertile as You Look?"

39. See *2015 Assisted Reproductive Technology National Summary*, 47.

40. Michael R. Soules, "The Story behind the American Society for Reproductive Medicine's Prevention of Infertility Campaign," *Fertility and Sterility* 80, no. 2 (August 2003): 295–99; Sora Song et al., "Making Time for a Baby," *Time* 159, no. 15 (April 15, 2002): 58.

41. Mark V. Sauer, Richard J. Paulson, and Rogerio A. Lobo, MD, "Reversing the Natural Decline in Human Fertility: An Extended Clinical Trial of Oocyte Donation to Women of Advanced Reproductive Age," *JAMA* 268, no. 10 (September 9, 1992): 1275–79, doi:10.1001/jama.1992.03490100073030; Mark V. Sauer, Richard J. Paulson, and Rogerio A. Lobo, "Pregnancy in Women 50 or More Years of Age: Outcomes of 22 Consecutively Established Pregnancies from Oocyte Donation," *Fertility and Sterility* 64, no. 1 (July 1995): 111–15.

42. In all, 18 percent of cycles in women aged forty-one and forty-two used donor eggs, 34 percent in those age forty-three and forty-four, and fully 71 percent of cycles in women over the age of forty-four. Just 8 percent of women age forty-one and forty-two gave birth using their own eggs, 3 percent at age forty-three and forty-four, and just 1 percent of women forty-four and older. See *2015 Assisted Reproductive Technology National Summary*, 47.

43. "Ordering Donor Eggs," Cryos, accessed October 8, 2018, https://usa.cryosinternational.com/donor-eggs/ordering-donor-eggs.

44. "Donor Fees and Costs," ConceiveAbilities, accessed November 28, 2017, https://www.conceiveabilities.com/parents/egg-donor-cost.

45. See the websites of Fertility Solutions, a West Coast agency that matches donors and recipients, http://www.fertility-solutions.com/egg-donors.html; Cryos International, a long-standing sperm bank that now offers frozen eggs, https://usa.cryosinternational.com/; and the Center for Reproductive Medicine, a prominent IVF practice in New York where recipients and donors remain anonymous, https://ivf.org/treatments-and-services/crms-pioneering-donor-egg-program/more-information-donor-egg.

46. See 1999 Results and *2015 Assisted Reproductive Technology National Summary*, 47. Comparisons between years are approximate because there were different ways of collecting and organizing data in earlier years, and those mechanisms often changed from year to year.

47. Margaret Marsh and Wanda Ronner, *The Fertility Doctor: John Rock and the Reproductive Revolution* (Baltimore: Johns Hopkins University Press, 2008), 297.

48. RAND Law and Health Initiative, *How Many Frozen Human Embryos Are Available for Research?*, Research Brief RB-9038 (Santa Monica, CA: RAND, 2013), https://www.rand.org/pubs/research_briefs/RB9038/index1.html; Tamar Lewin, "Industry's Growth Leads to Leftover Embryos, and Painful Choices," *New York Times*, June 17, 2015.

49. Jessica Cussins, "Embryos for Sale? When You Want Them, How You Want Them, or Your Money Back," *Psychology Today*, November 29, 2012, https://www.psychologytoday.com/blog/genetic-crossroads/201211/embryos-sale.

50. I. Glenn Cohen and Eli Adashi, "Made-to-Order Embryos for Sale—A Brave New World?," *New England Journal of Medicine* 368 (June 2013): 2517–19, http://www.nejm.org/doi/full/10.1056/NEJMsb1215894; Alan Zarembo, "An Ethics Debate over Embryos on the Cheap," *Los Angeles Times*, November 19, 2012, http://www.nejm.org/doi/full/10.1056/NEJMsb1215894.

51. Robert Klitzman, MD, and Mark V. Sauer, MD, "Creating and Selling Embryos for 'Donation': Ethical Challenges," *American Journal of Obstetrics & Gynecology* 212, no. 2 (February 2015): 167–70. Quote on p. 167. Sauer, readers will recall, had created embryos for his

patients in the 1990s, although his purpose was different. It appears that he had rethought the ethical issues since that time.

52. The website of California Conceptions provides no statistics. Zeringue's patient profile on the CDC site is tilted toward both women in their forties and the use of donor eggs; see "2015 Assisted Reproductive Technology Fertility Clinic Success Rates Report," Centers for Disease Control and Prevention, accessed October 8, 2018, https://www.cdc.gov/art/reports /2015/fertility-clinic.html.

53. See "Why Choose Snowflakes?," Snowflakes, accessed October 8, 2018, https://www .nightlight.org/snowflakes-embryo-adoption-donation/embryo-adoption/why-choose -snowflakes/.

54. See the NEDC website, accessed October 8, 2018, https://app.embryodonation.org/. For 2015 statistics for Keenan's practice, see "View ART Data," Centers for Disease Control and Prevention, accessed December 21, 2017, https://nccd.cdc.gov/drh_art/rdPage.aspx?rd Report=DRH_ART.ClinicInfo&ClinicId=436&ShowNational=0.

55. See *2015 Assisted Reproductive Technology National Summary*, 52.

56. Alex Kuczynski, "Her Body, My Baby," *New York Times Magazine*, November 28, 2008, http://www.nytimes.com/2008/11/30/magazine/30Surrogate-t.html?pagewanted=all&_r= commentsContainer.

57. Comments on Kuczynski, "Her Body, My Baby."

58. Kuczynski, "Her Body, My Baby."

59. Jane Ganahl, "Doug and Eric: Zing Went the Strings," *San Francisco Chronicle,* March 14, 2004, https://www.sfgate.com/living/article/LOVE-STORIES-Doug-and-Eric-Zing-went -the-2808523.php.

60. Mundy, *Everything Conceivable*, 134. Eric and Doug married in San Francisco in February 2004 after the city's mayor issued a ruling that same-sex marriages were legal. The ruling was challenged, and it wasn't until 2013 that same-sex marriage was legalized in the state. See Ganahl, "Doug and Eric."

61. This range is derived from viewing multiple surrogacy websites, some of which offer much more detailed cost breakdowns than others.

62. Mundy, *Everything Conceivable*, 134.

63. Mundy, *Everything Conceivable*, 146.

64. Jeannie Ralston, "I Gave Birth to Triplets for My Friend," *Ladies' Home Journal* (October 2009).

65. Information from Wanda Ronner. See also "Schizencephaly," US Department of Health and Human Services, accessed October 8, 2018, https://rarediseases.info.nih.gov/diseases/166 /schizencephaly.

66. Tamar Lewin, "Coming to U.S. for Baby, and Womb to Carry It," *New York Times*, July 6, 2014, 1, 12–13. Quote on p. 12.

67. See *2015 Assisted Reproductive Technology National Summary*, 5. Gestational carriers accounted for less than 1 percent of the 231,936 cycles performed that year, about the same percentage as in 1999, when gestational surrogates accounted for 821 cycles. See 1999 Results, 918. Of the 370 clinics, 360 reported their data to SART, and information is based on that reporting group.

68. Kiran M. Perkins, Sheree L. Boulet, Denise J. Jamieson, and Dmitry M. Kissin, "Trends and Outcomes of Gestational Surrogacy in the United States," *Fertility and Sterility* 106, no. 2 (August 2016): 437.

69. A surrogate is likelier to have her egg fertilized, so when one moves from "cycles initiated" to embryos transferred into the uterus, the pregnancy rates for surrogates are higher. When a report refers to a "non-donor" cycle, it generally involves the intended mother's egg and her male partner's sperm. A "donor" cycle uses a donor egg and presumably the sperm of an intended father. Same-sex male couples sometimes use sperm from both partners. In 2009, 2 percent of embryo transfers used gestational carriers, up from about 1.3 in 1999. Centers for Disease Control and Prevention, American Society for Reproductive Medicine, Society for Assisted Reproductive Technology, *2009 Assisted Reproductive Technology Success Rates: National Summary and Fertility Clinic Reports* (Atlanta, GA: US Department of Health and Human Services: 2011), 91. Data from slides, p. 3. Perkins et al., "Trends and Outcomes," 437.

70. Perkins et al., "Trends and Outcomes." Percentages compiled from table on p. 441. See *2015 Assisted Reproductive Technology National Summary*, 53.

71. See *2015 Assisted Reproductive Technology National Summary*, 5.

72. Lewin, "Coming to U.S. for Baby," 1.

73. Adam Liptak, "Supreme Court Ruling Makes Same-Sex Marriage a Right Nationwide," *New York Times*, June 26, 2015, https://www.nytimes.com/2015/06/27/us/supreme-court-same-sex-marriage.html.

74. Perkins et al., "Trends and Outcomes," 436. This figure may be low, however, since it depends on self-reporting to the clinic.

75. Gary J. Gates and Taylor N. T. Brown, "Marriage and Same-Sex Couples after *Obergefell*," Williams Institute, November 2015, https://williamsinstitute.law.ucla.edu/wp-content/uploads/Marriage-and-Same-sex-Couples-after-Obergefell-November-2015.pdf, 1, 4. Abbie E. Goldberg, Nanette K. Gartrell, and Gary Gates. "Research Report on LGB-Parent Families," Williams Institute, July 2014, https://williamsinstitute.law.ucla.edu/wp-content/uploads/lgb-parent-families-july-2014.pdf, 1, 2, highlights the paucity of research on gestational surrogacy among families with a lesbian, gay, or bisexual parent but suggests that what research does exist indicates that surrogacy is used primarily by affluent gay men.

76. Amanda Duberman, "I Went to an Egg Freezing Cocktail Party; Here's What I Learned," *Huffington Post*, August 20, 2014, https://www.huffingtonpost.com/2014/08/20/egg-freezing-party-egg-banxx_n_5688461.html.

77. Sarah Elizabeth Richards, *Motherhood Rescheduled: The New Frontier of Egg Freezing and the Women Who Tried It* (New York: Simon and Schuster, 2013), 241. There is no sign of a baby on the "About" section of her current website, accessed October 8, 2018, https://www.sarahelizabethrichards.com/about.

78. C. Versaci, G. Dani, M. Antinori, and H. A. Selman, "Successful Fertilization and Pregnancy after Injection of Frozen-Thawed Round Spermatids into Human Oocytes," *Human Reproduction* 3 (March 12, 1997): 554–56.

79. See "About," Oncofertility Consortium, accessed October 8, 2018, http://oncofertility.northwestern.edu/about-oncofertility-consortium.

80. See "ASRM: Egg Freezing No Longer 'Experimental' Technique," Oncofertility Consortium, October 19, 2012, http://oncofertility.northwestern.edu/blog/2012/10/asrm-egg-freezing-no-longer-experimental-technique.

81. Jon Bardin, "Freezing of Human Eggs No Longer an Experimental Procedure," *Los Angeles Times*, October 19, 2012, http://articles.latimes.com/2012/oct/19/news/la-heb-freezing-of-human-eggs-no-longer-experimental-20121019.

82. Richards, *Motherhood Rescheduled*, 142.

83. "More Women Are Freezing Their Eggs, but Will They Ever Use Them?," *Morning Edition*, November 24, 2015, https://www.npr.org/sections/health-shots/2015/11/24/456671203/more-women-are-freezing-their-eggs-but-will-they-ever-use-them. The exact number of women who froze their eggs in 2013 was 3,938. See the patient fact sheet on egg freezing: "Can I Freeze My Eggs to Use Later if I'm Not Sick?," ReproductiveFacts.org, accessed February 22, 2018, http://www.reproductivefacts.org/news-and-publications/patient-fact-sheets-and-booklets/documents/fact-sheets-and-info-booklets/can-i-freeze-my-eggs-to-use-later-if-im-not-sick/.

84. Richards, *Motherhood Rescheduled*, 144–46.

85. A total of 376 clinics had such websites. N. Joshi, S. B. Schon, P. Masson, and T. Zore, "Assessment of Fertility Clinic Websites on Oocyte Cryopreservation," *Fertility and Sterility* 108, no. 3, suppl. (September 2017): 189, http://dx.doi.org/10.1016/j.fertnstert.2017.07.559

86. Conversation with an anonymous physician.

87. A. H. Handyside, E. H. Kontogianni, K. Hardy, and R. M. L. Winston, "Pregnancies from Biopsied Human Preimplantation Embryos Sexed by Y-Specific DNA Amplification," *Nature* 344 (1990): 768; A. H. Handyside, J. G. Lesko, J. J. Tarín, R. M. L. Winston, and M. R. Hughes, "Birth of a Normal Girl after In Vitro Fertilization and Preimplantation Diagnostic Testing for Cystic Fibrosis," *New England Journal of Medicine* 327 (1992): 905–9.

88. See "PGD Conditions," Human Fertilisation and Embryology Authority, accessed February 20, 2018, https://www.hfea.gov.uk/pgd-conditions/.

89. See "2016 IVF Success Rates," Advanced Fertility Center of Chicago, accessed February 20, 2018, https://www.advancedfertility.com/pgs-ivf-genetic-testing.htm, for a good nontechnical description of both PGD and PGS.

90. Cathy Herbrand, "Three-Person IVF: What Makes Mitochondrial Donation Different?," January 11, 2016, http://www.bionews.org.uk/page_604026.asp.

91. Anne Claiborne, Rebecca English, and Jeffrey Kahn, eds., *Mitochondrial Replacement Techniques: Ethical, Social, and Policy Considerations* (Washington, DC: National Academies Press, 2016), xiv.

92. Margaret Marsh, "'Three Parent Embryos' in the News in 2016," IVF and Beyond, August 8, 2016, https://marshronner.rutgers.edu/2016/08/08/three-parent-embryos-in-the-news-in-2016/. Institute of Medicine, *Report in Brief: Mitochondrial Replacement Techniques: Ethical, Social, and Policy Considerations* (Washington, DC: National Academies of Sciences, Engineering, and Medicine, February 2016), 1. The PDF of the full report is available at http://www.nap.edu/21871.

93. The appropriations bill is discussed by Kurt R. Karst, "Senate and House Lawmakers Add to FDA's To-Do List in Fiscal Year 2016 Appropriations Bills," *FDA Law Blog*, July 24, 2015, http://www.fdalawblog.net/2015/07/senate-and-house-lawmakers-add-to-fdas-to-do-list-in-fiscal-year-2016-appropriations-bills/. See section 745 for the full text. On the FDA announcement, see Rob Stein, "Babies with Genes from Three People Could Be Ethical, Panel Says," *All Things Considered*, February 3, 2016, https://www.npr.org/sections/health-shots/2016/02/03/465319186/babies-with-genes-from-three-people-could-be-ethical-panel-says.

94. Ariana Eunjung Cha, "To Your Health: FDA Cracks Down on Company Marketing 'Three-Parent' Babies," *Washington Post*, August 8, 2017, https://www.washingtonpost.com/news/to-your-health/wp/2017/08/07/fda-cracks-down-on-company-marketing-three-parent-babies/?utm_term=.5cb57c71b9b0.

1. The medical definition of gross negligence is "the reckless provision of health care that is clearly below the standards of accepted medical practice, either without regard for the potential consequences, or with willful and wanton disregard for the rights and/or well-being of those for whom the duty is being performed." See "Gross Negligence," Free Dictionary, accessed October 8, 2018, https://medical-dictionary.thefreedictionary.com/gross+negligence.

2. Ashley Surdin, "Birth of Octuplets Stirs Ethical Concerns," *Washington Post*, February 4, 2009. Rong-Gong Lin II and Jessica Garrison, "California Medical Board Revokes License of 'Octomom' Doctor," *Los Angeles Times*, June 2, 2011, http://articles.latimes.com/2011/jun/02/local/la-me-0602-octomom-doctor-20110602.

3. Lin and Garrison, "California Medical Board Revokes License." Even before he lost his medical license in the Octomom case, Kamrava had been sued; he had also lied about his lab certification. And yet he remained in practice. It's not clear how large his infertility practice was, but the SART data indicated that his number of IVF patients was small, fewer than twenty cycles in the years we looked at. For the lawsuits and his generally low reputation among his colleagues, see Alison Stateman, "The Fertility Doctor behind the 'Octomom,' " *Time*, March 7, 2009, http://content.time.com/time/nation/article/0,8599,1883663,00.html.

4. Kim Christensen, "Doctor with Ties to Fertility Scandal Won't Be Extradited by Mexico," *Los Angeles Times*, April 1, 2011; Kimi Yoshino, "UC Irvine Fertility Scandal Isn't Over," *Los Angeles Times*, January 20, 2006.

5. A. Francine Coeytaux, Marcy Darnovsky, and Susan Berke Fogel, "Assisted Reproduction and Choice in the Biotech Age: Recommendations for a Way Forward," *Contraception Journal* (January 2011): 5. Judith Daar, "Federalizing Embryo Transfers: Taming the Wild West of Reproductive Medicine?," *Columbia Journal of Gender and Law* 23, no. 2 (2012): 1–6, disagrees with the idea that reproductive medicine is unregulated, noting the existence of professional regulation and state tort law. State regulation, however, by statute or tort law, is confusing and often contradictory. Without national policies, there is no coherent regulation.

6. US Department of Health and Human Services, Centers for Disease Control and Prevention, "Implementation of the Fertility Clinic Success Rate and Certification Act of 1992: Proposed Model Program for the Certification of Embryo Laboratories," *Federal Register* 63, no. 215 (November 6, 1998): 60,178–89.

7. Margaret Foster Riley with Richard A. Merrill, "Regulating Reproductive Genetics: A Review of American Bioethics Commissions and Comparison to the British Human Fertilisation and Embryology Authority," *Columbia Science and Technology Law Review* 6 (2005): 1–64.

8. Christos Coutifaris, interview with the authors, December 6, 2015.

9. Alan DeCherney, interview with the authors, November 2, 2015.

10. Section 745 of the law said that "none of the funds made available by this Act may be used to review or approve an application for an exemption for investigational use of a drug or biological product under section 505(I) of the Federal Food, Drug, and Cosmetic Act (21 U.S.C. 355(i)) or section 351(a)(3) of the Public Health Service Act (42 U.S.C. 262(a)(3)) in research in which a human embryo is intentionally created or modified to include a heritable genetic modification." See Kurt R. Karst, "Senate and House Lawmakers Add to FDA's To-Do List in Fiscal Year 2016 Appropriations Bills," *FDA Law Blog*, July 24, 2015, http://www.fda lawblog.net/2015/07/senate-and-house-lawmakers-add-to-fdas-to-do-list-in-fiscal-year-2016 -appropriations-bills/.

11. Centers for Disease Control and Prevention, *National Public Health Action Plan for the Detection, Prevention, and Management of Infertility* (Atlanta, GA: Centers for Disease Control and Prevention, June 2014), http://www.cdc.gov/reproductivehealth/infertility/pdf/drh_nap_final_508.pdf.

12. CDC, *National Public Health Action Plan*, 3.

13. CDC, *National Public Health Action Plan*, 9, 13–14. For the follow-up (2016) video-conferenced grand rounds, plus a link to the activities that have occurred to date (2018), see "Infertility and Public Health," Centers for Disease Control and Prevention, accessed March 8, 2018, https://www.cdc.gov/reproductivehealth/infertility/publichealth.htm.

14. Thomas Prentice, "Doctors Are Divided on Multiple Births," *Times of London*, April 2, 1984, 3.

15. Information gathered by Wanda Ronner in extensive conversations with Eli Adashi, including one interview on October 2, 2017.

16. Debra Nussbaum, "Triplet Nation," *New York Times*, October 23, 2005, p. NJ1.

17. Suzanne Sanchez, "My Triplets Were Inseparable, Whatever the Risks," *New York Times*, March 4, 2007, 18.

18. "Multiple Pregnancy," American College of Obstetricians and Gynecologists, July 2015, https://www.acog.org/Patients/FAQs/Multiple-Pregnancy.

19. See notes to table A1 for sources.

20. See the Sperm Donor Catalogue available at the website of the Seattle Sperm Bank / European Sperm Bank USA, accessed October 8, 2018, https://www.europeanspermbank.com/en/.

21. Rene Almeling, *Sex Cells: The Medical Market for Eggs and Sperm* (Berkeley: University of California Press, 2011), quotes from 169–70.

22. For donor insemination for unmarried women in the 1930s through the 1950s, see Margaret Marsh and Wanda Ronner, *The Empty Cradle: Infertility in American from Colonial Times to the Present* (Baltimore: Johns Hopkins University Press, 1996), 165. See also Kara Swanson, *Banking on the Body: The Market in Blood, Milk, and Sperm in Modern America* (Cambridge, MA: Harvard University Press, 2014), chap. 6; Bridget Gurtler, "Synthetic Conception: Artificial Insemination and the Transformation of Reproduction and Family in Nineteenth and Twentieth Century America" (PhD diss., Rutgers University, 2013), chap. 5.

23. "Father Pleads Guilty to Killing Baby Born to Surrogate Mother," *Los Angeles Times*, August 8, 1995, http://articles.latimes.com/1995-08-08/news/mn-32653_1_surrogate-mother; David Whiting, "Surrogate Mom Fears for Triplets after Allegations of Abuse by Father," *Orange County Register*, September 20, 2017, https://www.ocregister.com/2017/09/20/surrogate-mom-fears-for-triplets-after-allegations-of-abuse-by-father/; Kelly Puente, "Woodland Hills Surrogate Mom Loses Custody Battle for Triplets," *Orange County Register*, January 16, 2018, https://www.ocregister.com/2018/01/16/woodland-hills-surrogate-mom-loses-custody-battle-for-triplets/.

24. Liza Mundy, *Everything Conceivable: How the Science of Assisted Reproduction Is Changing Our World* (New York: Anchor Books, 2008), 136.

25. Roli Srivastava, "Which Countries Allow Commercial Surrogacy?," *Reuters*, January 19, 2017, https://www.reuters.com/article/us-india-women-surrogacy-factbox/fatcbox-which-countries-allow-commercial-surrogacy-idUSKBN1530FP. Canada is considering decriminalizing compensated surrogacy, but as of this writing it has not done so.

26. "Thailand Bans Foreigners, Same-Sex Couples from Seeking Surrogacy Services," LGBTQ Nation, February 21, 2015, https://www.lgbtqnation.com/2015/02/thailand-bans-foreigners-same-sex-couples-from-seeking-surrogacy-services/; Charles Riley and Medhavi Arora, "In-

dia Wants to Stop Foreigners Paying Women to Have Babies," CNN, August 25, 2016, http://money.cnn.com/2016/08/25/news/india-surrogacy-foreigners/index.html; Tamar Lewin, "Coming to U.S. for Baby, and Womb to Carry It: Foreign Couples Heading to America for Surrogate Pregnancies," *New York Times*, July 5, 2015, https://www.nytimes.com/2014/07/06/us/foreign-couples-heading-to-america-for-surrogate-pregnancies.html.

27. Owen Bowcott, "Surrogacy Review to Tackle Laws Declared Unfit for Purpose," *The Guardian*, May 4, 2018, https://www.theguardian.com/lifeandstyle/2018/may/04/surrogacy-review-to-tackle-laws-declared-unfit-for-purpose. For a British surrogacy agency, see the website of Surrogacy UK, accessed October 8, 2018, http://www.surrogacyuk.org/.

28. For an example, see "About Orchids," a talk by scientist Jeff Nisker, who helped develop PGD in Canada, on the Tarrytown Meetings website, accessed October 8, 2018, http://thetarrytownmeetings.org/notes/about-orchids-jeff-nisker.

29. Quote from Jocelyn Kaiser, "U.S. Panel Gives Yellow Light to Human Embryo Editing," *Science*, February 14, 2017, http://www.sciencemag.org/news/2017/02/us-panel-gives-yellow-light-human-embryo-editing; Press release from the National Academies of Science, February 14, 2017, http://www8.nationalacademies.org/onpinews/newsitem.aspx?RecordID=24623. For the full report, see National Academies of Sciences, Engineering, and Medicine, *Human Genome Editing: Science, Ethics, and Governance* (Washington, DC: National Academies Press, 2017), https://doi.org/10.17226/24623.

30. Robert Debbs, conversation with the authors, February 2, 2018.

31. In America's IVF clinics in 2015, just 1 percent of patients forty-five and older were able to have a baby using their own eggs, and the odds were not very good for younger forty-somethings. They were 3 percent for women aged forty-three and forty-four, and just 8 percent for women age forty-one and forty-two.

32. Mark Sauer, "Reproduction at an Advanced Maternal Age and Maternal Health," *Fertility and Sterility* 103, no. 5 (May 2015): 1136–43. The pregnancies he discussed were achieved with donor eggs.

33. Joyce A. Martin, Brady E. Hamilton, Michelle J. K. Osterman, Anne K. Driscoll, and T. J. Mathews, "Births: Final Data for 2015," *National Vital Statistics Reports* 66, no. 1 (January , 2017): 2–5, https://www.cdc.gov/nchs/data/nvsr/nvsr66/nvsr66_01.pdf.

34. Marcia C. Inhorn, "Women, Consider Freezing Your Eggs," CNN, April 9, 2013, https://www.cnn.com/2013/04/09/opinion/inhorn-egg-freezing/index.html.

35. Alana Cattapan, Kathleen Hammond, Jennie Haw, and Lesley A. Tarasoff, "Breaking the Ice: Young Feminist Scholars of Reproductive Politics Reflect on Egg Freezing," *International Journal of Feminist Approaches to Bioethics* 7, no. 2 (Fall 2014): 236–47. Quote on p. 239.

36. Lynne M. Morgan and Janelle S. Taylor, "Op-Ed: Egg Freezing: WTF?* (*Why's This Feminist?)," *Feminist Wire*, April 14, 2013, http://thefeministwire.com/2013/04/op-ed-egg-freezing-wtf/.

37. Sonam Sheth, Shayanne Gal, and Skye Gould, "5 Charts Show How Much More Men Make Than Women," *Business Insider*, August 27, 2018, http://www.businessinsider.com/gender-wage-pay-gap-charts-2017-3.

38. Reniqua Allen, "Is Egg Freezing Only for White Women?," *New York Times*, May 22, 2016, 5.

39. Michelle Obama, *Becoming* (New York: Crown, 2018), 187–89.

40. Obama, *Becoming*, 189.

41. Obama, *Becoming*, 189–90, 199.

42. Barron S. Lerner, *The Breast Cancer Wars: Hope, Fear, and the Pursuit of a Cure in Twentieth-Century America* (New York: Oxford University Press, 2001), 173–74.

43. David B. Seifer, Linda M. Frazier, and David A. Grainger, "Disparity in Assisted Reproductive Technologies Outcomes in Black Women Compared with White Women," *Fertility and Sterility* 90, no. 5 (November 2008): 1701–10. The actual percentage was 4.6; see Dana B. McQueen, Ann Schufreider, Sang Mee Lee, Eve C. Feinberg, and Meike L. Uhler, "Racial Disparities in In Vitro Fertilization Outcomes," *Fertility and Sterility* 104, no. 2 (August 2015): 398–402. (Actual figures were as follows: 13.4 percent of the patients were Asian, 7.1 percent were Hispanic, and 74.2 percent were white.) See also Tarun Jain, "Socioeconomic and Racial Disparities among Infertility Patients Seeking Care," *Fertility and Sterility* 85, no. 4 (April 2006): 876–81. SART asks clinics to report the race of their patients, but those data are not publicly available. In addition, apparently only about 65 percent of clinics report such information. Our data come from studies done by medical researchers who were able to secure access to the SART data for particular anonymized studies or from reports of studies conducted at particular clinics. We would like to see such information included in the SART public information.

44. Ethics Committee of the American Society for Reproductive Medicine, "Disparities in Access to Effective Treatment for Infertility in the United States: An Ethics Committee Opinion," *Fertility and Sterility* 104, no. 5 (November 2015): 1104–10.

45. Ethics Committee, "Disparities in Access," 1106.

46. "Scholarship FAQ," International Council on Infertility Information Dissemination, Inc., accessed October 8, 2018, https://www.inciid.org/ivf-scholarship.

47. Christopher Herndon, Yanett Anaya, Martha Noel, Hakan Cakmak, and Marcelle I. Cedars, "Outcomes from a University-Based Low-Cost In Vitro Fertilization Program Providing Access to Care for a Low-Resource Socioculturally Diverse Urban Community," *Fertility and Sterility* 108, no. 4 (October 2017): 642–49.

48. Ethics Committee, "Disparities in Access," 1107.

49. Ethics Committee, "Disparities in Access," 1106; Herndon et al., "Outcomes," 647.

50. Giuliano Testa et al., "First Live Birth after Uterus Transplantation in the United States," *American Journal of Transplantation* 18:5 (May 2018): 1270–1274. Baylor had a second birth in March of 2018: http://news.bswhealth.com/releases/second-mother-who-received-transplanted-uterus-gives-birth?query=uterus. On the program at the University of Pennsylvania, Christos Coutifaris, personal conversation, March 19, 2018.

51. A healthy non-smoker can donate part of a lung, but otherwise, lung transplants generally come from a cadaveric donor: https://www.hopkinsmedicine.org/healthlibrary/test_procedures/pulmonary/lung_transplant_92,p07752

52. Dani Ejzenberg, Wellington Andraus, et al., "Livebirth after Uterus Transplant from a Deceased Donor in a Recipient with Uterine Infertility," *The Lancet* (December 4, 2018), https://doi.org/10.1016/S0140-6736(18)31766-5. We used lay terminology in the text to explain how the donor died. Physicians would say her death resulted from a subarachnoid hemorrhage.

53. See, for example, "Paving a New Path to Parenthood: Penn Medicine Launches First Clinical Trial for Uterine Transplant in the Northeast," *Penn Medicine News*, November 7, 2017, https://www.pennmedicine.org/cancer/pr-news/news-releases/2017/november/penn-medicine-launches-first-clinical-trial-for-uterine-transplant-in-the-northeast. On the likely cost, see Denise Grady, "Woman with Transplanted Uterus Gives Birth, the First in the U.S.,"

New York Times, December 2, 2017, https://www.nytimes.com/2017/12/02/health/uterus
-transplant-baby.html.

54. Karen Weintraub, "First Successful Uterus Transplant from Deceased Donor Leads to
Healthy Baby," *Scientific American* (December 5, 2018), https://www.scientificamerican.com
/article/first-successful-uterus-transplant-from-deceased-donor-leads-to-healthy-baby/.
Here, the reporter is paraphrasing Dr. O'Neill.

55. See Emily A. Partridge et al., "An Extra-Uterine System to Physiologically Support the
Extreme Premature Lamb," *Nature Communications* 8 (2017): 15112, https://www.nature.com
/articles/ncomms15112; "A Unique Womb-Like Device Could Reduce Mortality and Disability
for Extremely Premature Babies," CHOP News, April 25, 2017, http://www.chop.edu/news
/unique-womb-device-could-reduce-mortality-and-disability-extremely-premature-babies.

56. Howard Jones's talk is available at "Howard W. Jones, Jr., M.D. Birthday Celebration,"
video uploaded by ASRM, 49:06, https://vimeo.com/channels/asrm/17418251.

57. Debora Spar, *The Baby Business: How Money, Science, and Politics Drive the Commerce of
Conception* (Cambridge, MA: Harvard Business Review Press, 2006), 224. Emphasis is hers.

58. See HFEA's official website, accessed October 8, 2018, https://www.hfea.gov.uk/.

59. Eli Adashi and Glenn Cohen, "Preventing Mitochondrial Disease: A Path Forward,"
Obstetrics and Gynecology 131, no. 3 (March 2018): 553–55.

INDEX

Page numbers in *italics* indicate figures and tables.

abortion: anti-abortion "embryo adoption" programs, 163–64; legalization of, 7, 48; opposition to, 55; Republican party and, 106–7
academic medical center clinics, 127, 149, 150–52. *See also specific clinics and specific universities*
access to reproductive technologies, 8–9, 112, 127, 129–30, 189–90, 203–8. *See also* insurance coverage
Adashi, Eli, 162, 191, 192, 193, 211–12
adoption: costs of, 148; insurance coverage for, 113; as option for infertility, 13, 111
"adoption" of embryos, 163–64
Affordable Care Act, 207, 211
African Americans: access to ART and, 4–5, 129–30, 203–8; infertility and, 105; as patients of Daniell, 51–52; at Penn IVF program, 82; as surrogates, 136; wage gap and, 202
age: of donor egg recipients, 132; fertility and, 158–59, 200–201; infertility and, 105, 146–48, 155, 156; of patients, 94; pregnancy and, 199–202; use of ART by, *214*; of women at first birth, 104, 129
Albert Einstein Hospital, 89
Allen, Edgar, 16
Allen, Reniqua, 203

Almeling, Rene, 132, 137, 194
altruistic surrogacy, 133, 196–97, 198
American College of Obstetricians and Gynecologists, 144
American Fertility Society: guidelines of, 91–92, 110, 116; IVF and, 81; G. Jones and, 57; on regulation, 96; Wyden bill and, 118. *See also* American Society of Reproductive Medicine
American Medical Association, 47
American Society of Reproductive Medicine (ASRM): donated embryos and, 164; guidelines of, 162–63; insurance coverage and, 206–7; multiple births and, 193; NABER and, 144; on oocyte cryopreservation, 171, 172; public service campaign of, 159; voluntary mechanisms of, 186–87. *See also* American Fertility Society
Andrews, Lori, 112
Andrews, Mason, 55–56, 59, 60, 61, 62–63
Annas, George, 112
anti-abortion "embryo adoption" programs, 163–64
anti-feminism, 71, 72, 104–5, 107, 199
Arehart-Treichel, Joan, 38
Arrowsmith (Lewis), 18
artificial insemination, 117, 134, 139–40, 141, 195
Asch, Ricardo, 121–26, 136, 184
Asian Americans and access to ART, 4–5, 203–8
Asians and surrogacy services, 169, 197

ASRM. *See* American Society of Reproductive Medicine

assisted reproductive technology (ART): access to, 4–5, 8–9, 112, 127, 129–30, 189–90, 203–8; as business, 127, 149–53; controversy surrounding, 5–7; cultural significance of, 136–37, 154; definition of, x, 2; demographics of, 4–5; fertile couples and, 173–77; policy on, 7–8; pregnancies and births as result of, xi, 144–45, *216*; supply of and demand for, 153–57; US as Wild West of, 7, 10; use of, by age, *214*. *See also* costs; in vitro fertilization

Australia. *See* Melbourne IVF team

baby boom generation: infertility rates among, 105–6; origins of, 26; as parents, 104; pronatalism and, 43–44, 45, 71

Baby M case, 134

Balmaceda, Jose, 122, 123–24, 126

Bartlett, Marshall, 17

Beatty, Alan, 35

Beckworth, John, 105

Berkowitz, Gertrud, 105

Bevis, Douglas, 41–42

Bird, Robert, 23–24, 26

Birnbaum, Frieda, 158

Blasco, Luis, 81

blastocysts, 38, 49

blood tests in infertility workup, 3

Boff, Shannon, 135

Boncompagni, Tatiana, 146

Boston Women's Health Collective, 44

Brave New World (Huxley), 19, 20, 23, 24, 210

Brickman, Edith, 105

Brody, Jane, 30

Brown, John, 74–75

Brown, Lesley, 31–32, 50, 54, 74–75

Brown, Louise: birth of, 31; R. Edwards, Steptoe, and, *98*; media attention to, x, 50, 74; reaction to birth of, 8, 54–56

Bush, George W., 163

business model: of ART, 127; of fertility services, 122, 149–53; of Jones Institute, 76–77; of Marrs, 77

Butler, Kathy and Gary, 138, 139

CAH (congenital adrenal hyperplasia), 57–58

Califano, Joseph, 54–55, 63, 64–65

California, IVF in, 67–70

California Conceptions, 163

Calvert, Crispina and Mark, 136

Caplan, Arthur, 121, 133

Carr, Elizabeth Jordan, 53, 62, 74

Carr, Judith, 53, 74, 75–76

Carter, Jimmy, 54, 55, 71

Carter, Tim Lee, 54

Catholic Church in Philadelphia, 85–86

CCRM (Colorado Center for Reproductive Medicine), 153

Center for Reproductive Health, University of California, San Francisco, 206, 207–8

Center for Reproductive Medicine, Weill Cornell Medical College, 152

Centers for Disease Control and Prevention (CDC), *National Public Health Action Plan*, 188–90, 193, 211

Chadwick, James, 35

Challender, John and Debbie, 125–26

Chang, M. C., 36

Chavkin, Wendy, 112

chemical pregnancy, 40–41

childbearing, trend toward later, 146–48, 155–56, 158–59, 199–202. *See also* pregnancy

chromosome abnormalities, screening for, 174

Clinton, Bill, 143, 144

Clomid, 204, 227n7

clomiphene citrate, 121–22

Cohen, Glenn, 162, 211–12

Cohen, Jacques, 83–84, 141–42

Cohen, Richard, 60

Colorado Center for Reproductive Medicine (CCRM), 153

Columbia-Presbyterian Hospital, 42, 50, 138–39

Columbia University, 127

commercialization of egg donation, 131–32

commodification of eggs, sperm, and embryos, 193–98

community hospitals, IVF programs at, 85–87

ConceiveAbilities, 160

Coney, PonJola, 86–87, 89, 93, *101*

congenital adrenal hyperplasia (CAH), 57–58

congressional hearings: on consumer protection, 116–17; on IVF, 54, 92, 108; on reproductive technologies, 109–13; on Schroeder
bill, 113, 114
conservatives/conservatism: pro-family, 71–73;
reproductive technologies and, 45–48, 106–8
consumer protection and reproductive services, 116–19, 122, 128, 186
Cooley, Denton, 68
Coontz, Stephanie, 156
Corner, George, 16, 22
Corson, Stephen, 88, 89
costs: of adoption, 148; consumer protection
and, 116–17; of donor eggs, 160; of egg
freezing and storage, 203; of gestational surrogacy, 166; of IVF, 62, 76, 82, 130; of multiple births, 192; of reproductive technologies, 114, 152; socioeconomic status and,
205–6; of uterus transplants, 209
Coutifaris, Christos, 151, 187–88, 208
CRISPR/Cas 9, 198–99
cryopreservation: complexities of, 170–73; egg
banks and, 160; elective or "social," 157, 170,
171–72, 187, 201–2; growth of, 120; history
of, 137–38; for medical reasons, 171, 172;
pregnancies after, 95; race and, 203
Cryos USA, 160, 194
culture: gender roles, parenthood, and, 107,
156; infertility and, 43–45; significance of
assisted reproductive technologies within,
136–37, 154; of wartime prosperity, 24–25.
See also baby boom generation; political culture
cytoplasm, 141–42

Dalkon Shield, 73
Daniell, James, 51–52
Dean, Charles, Jr., 60
DeBakey, Michael, 68
Debbs, Robert, 199
DeCherney, Alan: on business model, 127; on
career, 150–51; on fertility and age, 105; on
for-profit clinics, 92; on misappropriation of
eggs and embryos, 124–25; patients of, 82;
on surgical skills, 84; on tubal surgeons, 67;
Yale clinic of, 70, 80

Del Zio, Doris, 42, 50
demand for IVF, 61–62
development of IVF in US: at community hospitals, 85–87; evolution of, 79; at medical
schools, 80–83; at private clinics, 90–93
diagnosis: of patients, 94; preimplantation genetic, 156–57, 173–74, 198
Dickey-Wicker amendment, 143–44, 176, 188
direct marketing of eggs, 160–61
DNA: editing, 198–99; mitochondrial, 174–
75; nuclear, 142, 174
Doheny, Kathleen, 132
Doisy, Edward, 16
donor eggs: acquisition of, 160–61; compensation for, 194, 195; concerns expressed about,
148; cytoplasm from, 141–42; demand for,
159–60; embryos created with, 138; main-
streaming of, 148–49; parental rights and,
137; postmenopausal childbearing and,
158–59; retrieval of, 90; sources of, 131–32;
use of, 94–96, 129, 164
donor embryos, 94–96, 138, 161–64
donor insemination, 139–40, 141, 243n24,
251n76
Down syndrome, 174
Dr. Kershaw's Cottage Hospital, 48–49, 79
Duberman, Amanda, 170

Eastern Virginia Medical School, 55–56, 60,
61, 63, 66, 79. See also Jones Institute
Eckhardt, Vicki and Bill, 116
economic issues: commodification of eggs,
sperm, and embryos, 193–98; funding for
embryo research, 7, 52, 63–67, 72–73, 115,
143–44. See also costs
ectopic pregnancy, 2, 49
editing DNA, 198–99
Edwards, Robert: at Bourn Hall, 79, 99; Louise
Brown and, 98; criticism of, at medical ethics forum, 46–47; embryo transfer and, 35–
36; G. and H. Jones and, 30, 58–59; life and
career of, 34–35, 36–37; Medical Research
Council and, 39–40; multiple births after
IVF and, 191; Nobel Prize for, 51; Purdy and,
39; Steptoe and, 30, 32–33, 34, 37, 48–49, 99.
See also Brown, Louise

egg banks, 160, 194
Eggbanxx, 170, 172
egg-freezing programs. *See* cryopreservation
eggs: commodification of, 193–98; direct marketing of, 160–61; fertilization of, 37–39, *97*; misappropriation of, 124–25; retrieval of, 1, 4, 89–90; transfers of, 254n24; vitrification of, 170–71. *See also* donor eggs; GIFT
egg-sharing programs, 94–96
Eisenberg, Esther, 83, 86, 87
embryo research: conservative opposition to, 48, 108; federal funding for, 7, 52, 63–67, 72–73, 115, 143–44; on primates, 39, 47, 80–81; of Rock and Hertig, 17; stem cells and, 188
embryos: blastocysts, 38, 49; diagnosing, screening, and altering, 198–99; discarding, 63; from donors, 94–96, 138, 161–64; misappropriation of, 121–26; premade, 138–39, 194; sale of, 194–95; storage of, 161; "three-parent," 141, 185; transfer of, 35–36, 168–69, 191, 193. *See also* cryopreservation; embryo research
Empty Cradle, The (Marsh and Ronner), ix
endometriosis, 94
endometrium, 17
ethics: of creation and sale of embryos, 139, 162, 194–95; of R. Edwards, 46–47; of fertility clinics, 122–23; of MRT, 175; National Advisory Board on Ethics in Reproduction, 144; scandals and, 184–85; of surrogacy, 133, 136, 197–98; violations of, 121. *See also* access to reproductive technologies
Ethics Advisory Board (HEW), 48, 52, 54–55, 63–65, 66, 190
Ethington, Eric, 165–66
Extend Fertility, 171, 172

fallopian tubes, infertility due to blocked or damaged, 12–13, 73–77
Faludi, Susan, 104
families: characteristics of, 9; gender roles in, 107, 156; idea of, as changing/contested, 6–7, 43, 71, 72, 104, 119, 185, 186; IVF as decoupling pregnancy and genetic parenthood, x–xi, 46. *See also* motherhood; same-sex couples

FDA. *See* Food and Drug Administration
Federal Employee Family Building Act, hearings on, 113, 114
feminism: backlash against, 104; and elective egg freezing, 201–2; and medicalization, 45; and "motherhood mandate," 119; and multiple births, 192; opposition to ART and, 45, 107; and pro-choice Republicans, 107; reproductive technologies and, 45, 119; second-wave, 71
fertility and age, 158–59, 200–201. *See also* infertility
Fertility Center of Illinois, 205
Fertility Clinic Success Rate and Certification Act of 1992 (Wyden Act), 118–19, 122, 128, 186
Fertility Doctor, The (Marsh and Ronner), ix
fertility drugs and multiple births, 193
fertility services: consumer protection and, 116–19, 122, 128, 186; as industry, 122, 149–53; insurance coverage for, 185, 206–7; as part of basic health care, 207; for same-sex couples, 157
fertilization: criteria for, 37–39; photograph of, *97*. *See also* GIFT; in vitro fertilization
fetal reduction, 191, 192
FINRRAGE, 107
Firestone, Shulamith, *The Dialectic of Sex,* 45
Fleming, Anne Taylor, 61
Fletcher, Joseph, 47
Food and Drug Administration (FDA): genetic modification and, 142–43; MRT and, 174–75, 176–77
Ford, Betty, 204
foreign nationals and surrogacy, 169, 197
for-profit clinics, 90–93, 127–28, 150, 151–53. *See also specific clinics*
Fowler, Ruth, 36, 37
Franklin, Robert, 68
Free Hospital for Women, 11, 16
Friedrich, Otto, 102
frozen egg market, 194. *See also* cryopreservation
gamete intrafallopian transfer (GIFT), 89, 122, 125, 130, 254n24
Garcia, Celso-Ramon, 80, 84

Index

Gates, Alan, 36
GayBINGO, 195
gender roles and parenthood, 107, 156
genetic modification, 142–43, 175–76
gestational surrogacy: altruistic, 133, 196–97,
    198; attempt to criminalize, 114; availability
    of, 195; compensation for, 196–98; com-
    plexities of, 164–70; expansion of IVF and,
    94; first program for, 96; history of, 135–36;
    international, 197; H. Jones on, 210; markets
    for, 196; parental rights and, 136–37; same-
    sex couples and, 195; "traditional," 133–34;
    transactional behavior in, 196; use of, 133;
    Utian and, 134–35
GIFT (gamete intrafallopian transfer), 89,
    122, 125, 130, 254n24
Glenister, T. W., 37
gonadotropins, 4, 57, 227n7
Goodman, Ellen, 60–61
Gore, Al, 108, 109, 110, 114, 115
Grayson, Sarajean, 157–58
Great Britain: Bourn Hall, 79, 99, 135; consul-
    tant physicians in, 232n7; Human Fertilisa-
    tion and Embryology Authority, 8, 9, 109,
    174, 211; infertility research in, 40; laparo-
    scopic surgery in, 34; Medical Research
    Council, 39–40; MRT in, 174; policy devel-
    opment in, 7–8; regulation of reproductive
    technologies in, 108–9; surrogacy law in,
    136, 197
Green, Ronald, 143
Greenblatt, Robert, 121–22
Greene, Jody, 68, 69, 70
Grills, Dianne, 51
Grobstein, Clifford, 66, 109
gross negligence, definition of, 260n1
Growing Generations, 166
GynCor, 127

Hamilton, W. J., 37
Handyside, Alan, 174
Harris, Patricia, 65
Hartman, Carl, 22–23
Hastert, Dennis, 112
Health, Education, and Welfare (US, HEW),
    Ethics Advisory Board, 48, 52, 54–55, 63–
    65, 66, 190

Heape, Walter, 18
Henig, Robin Marantz, 42
heritable disease, preimplantation diagnosis of,
    156–57, 173–74, 198
Hertig, Arthur, 17
Heuser, Chester, 22
Hilling, Cathy, 164, 196
Hispanics: access to ART and, 203–8; wage
    gap and, 202
Human Embryo Research Panel, 143
Human Fertilisation and Embryology Author-
    ity (UK), 8, 9, 109, 174, 211
Huppert, Leonore, 149
Huxley, Aldous, Brave New World, 19, 20, 23,
    24, 210
Hyde, Henry, 114
Hyde Amendment, 55
hysterosalpingograms, 3

ICSI (intracytoplasmic sperm injection), 4,
    139, 140–41
"impaired fecundity" category, 153
INDs (investigational new drugs), 176
infertility: age and, 105, 146–48, 155, 156;
    cultural climate and, 43–45; as epidemic,
    102; evaluations of, 3; medical help for, 106,
    129, 149–50; as national public health prob-
    lem, 188–90; rates of, 71, 105, 153; research
    on, in UK, 40; success rates of treatment of,
    154–55; from tubal disease, 12–13, 73–77;
    unexplained, 203–4; uterine factor, 209; of
    women, claims about, 102–3, 104–5. See
    also insurance coverage; male infertility
Inhorn, Marcia, 201–2
insemination: artificial, 117, 134, 139–40, 141,
    195; donor, 243n24, 251n76; intrauterine,
    228n7; subzonal, 140
Institute of Medicine and MRT, 174–75, 176
instruments used for IVF, 84
insurance coverage: for infertility services and
    assisted reproduction, 185, 206–7; for IVF,
    113, 114, 115, 130–31, 204, 220
International Council on Infertility Informa-
    tion Dissemination, 206
international surrogacy, 197
intracytoplasmic sperm injection (ICSI), 4,
    139, 140–41

Index

intrauterine devices (IUDs), 73
intrauterine insemination, 228n7
investigational new drugs (INDs), 176
in vitro fertilization (IVF): announcement of,
    23–24; in Australia, 40–41; as decoupling
    pregnancy and genetic parenthood, x–xi;
    demographics of, 4–5; donor mtDNA use
    in, 174–75; R. Edwards, Steptoe, and, 30,
    37–38; first baby conceived using, 31; growth
    and expansion of, 128–31; media coverage of,
    23–25, 26–28, 31–32, 40; Menkin and, 21–
    22; multiple births after, 190–93, *218*; patient
    experience of, 1–2, 3–4; Pincus and, 19–20;
    race and, 203–4; Rock, Menkin, and, 11–12,
    13–14, 18, 22–24, 25, 37–38; skepticism of,
    22–23; as term, 2–3; Vatican and, 29. *See also*
    Brown, Louise
IUDs (intrauterine devices), 73
IVF. *See* in vitro fertilization

Jacobson, Cecil, 116, 184
Johnson, Anna, 136
Johnson, Martin, 34, 35, 36, 40
Johnston, Ian, 69–70, 75, *99*
Johnston, Jillian Elizabeth, 83
Jones, Christy, 171
Jones, Georgeanna Seegar, *99, 100, 101*; busi-
    ness model of, 76–77; career of, 53, 55, 56–
    57, 58, 66; at Eastern Virginia Medical
    School, 59–63; R. Edwards and, 30, 47, 58–
    59; stimulated cycles and, 62; Wentz and, 70
Jones, Howard W., Jr., *99, 100, 101*; on artificial
    wombs, 210; business model of, 76–77; ca-
    reer of, 53, 55, 56–58, 66; at Eastern Virginia
    Medical School, 59–63; R. Edwards and, 30,
    47, 58–59; number of embryos transferred
    and, 191; on regulation, 109
Jones Institute, 95, 131, 142, 151. *See also* East-
    ern Virginia Medical School
Jonsen, Albert R., 133

Kamrava, Michael, 184, 187
Kass, Leon, 46
Katz, Jay, 133
Keane, Noel, 135
Keenan, Jeffrey, 163, 164

Kenley, James, 60
Kennedy, Anthony, 169
Klein, Renate, 107
Kletsky, Oscar, 68, 69
Klitzman, Robert, 162
Kolata, Gina, 138, 139
Krol, John Cardinal, 85
Kuczynski, Alex, 164–65, 196

Ladies' Home Journal, 148–49
Landers, Ann, 43
Lang, Susan, 105
laparoscopic surgery, 33, 34, 84
Laurence, William, 19
Leeton, John, 40, 95, *99*
legislation: about surrogacy in Great Britain,
    136, 197; Dickey-Wicker amendment, 143–
    44, 176, 188; Fertility Clinic Success Rate
    and Certification Act of 1992, 118–19, 122,
    128, 186; Hyde Amendment, 55
Lenz, Susan, 90
Lewis, Sinclair, *Arrowsmith,* 18
Lobo, Rogerio, 132, 200
Loeb, Jacques, 18–19
Lopata, Alex, 70, *99*
Luken, Thomas, 114
luteal phase defect, 57
luteinizing hormone, 50

MacDonald, Alastair, 50
MacPhail, Neil, 15
male infertility: donor sperm for, 139–40;
    ICSI for, 4, 139, 140–41; IVF for, 88, 89,
    94, 140
malpractice, 184
Markens, Susan, 136
marriage, trend toward later, 156
Marrs, Richard: on academic medical centers,
    151–52; business model of, 77; on donor
    egg patients, 131; education and career of, 5,
    67–68; on equipment used for IVF, 84; fro-
    zen embryos and, 95, 137; patients of, 82,
    93; on regulation, 92–93; research of, 69–70
Marx, Jean, 38
Mastroianni, Luigi: career of, 80–81; Catholic
    Church and, 85–86; DeCherney and, 70,

150–51; donor eggs and, 96; IVF program of, 78–79; laparoscopy and, 84; legacy of, 151; Rock and, 26, 80

McCarthy-Keith, Desiree, 203

McCormick, Richard A., 64

McKusick, Victor, 58

media coverage: of birth of Louise Brown, x, 50, 74; of IVF, 23–25, 26–28, 31–32, 40; Rock and, 27–28

Melbourne IVF team: chemical pregnancy of, 49; donor eggs and, 95; fertilization success of, 41; Marrs and, 69–70; patients of, 75; stimulated cycles and, 62; success of, 79

Menkin, Miriam, *98*; career of, 6; IVF and, 11–12, 17, 21–22, 37–38; Rock and, 18, 20–21, 29

Menkin, Valy, *97*

Mishell, Daniel, 69

mitochondrial DNA (mtDNA), 174–75

mitochondrial replacement techniques (MRT), 174–77, 198, 211–12

"mommy penalty," 202

moral objections to reproductive technologies, 46–47. *See also* ethics

Morgan, Lynne, 202

Morrison, Bruce, 111, 112–13, 115

motherhood: changing definitions of, 127–28; postponement of, 146–48, 155–56, 158–59, 199–202

MRT (mitochondrial replacement techniques), 174–77, 198, 211–12

mtDNA (mitochondrial DNA), 174–75

Mt. Sinai Hospital, Cleveland, 96, 134–35, 169

Mulligan, William, 28

multiple births after IVF, 190–93, *218*

Mundy, Liza, 147, 166

Naftolin, Frederick, 63

National Advisory Board on Ethics in Reproduction, 144

National Bioethics Advisory Commission, 144

National Commission for the Study of Ethical Problems in Medicine and Biomedical and Behavioral Research, 66

National Embryo Donation Center, 163–64

National Institutes of Health (NIH): Asch investigation and, 124–25; funding ban and, 7, 73, 187–88; Human Embryo Research Panel, 143; Soupart and, 51

National Organization of Non-Parents, 43

National Survey of Family Growth, 253–54n20

Navas, Marna, 131

nDNA (nuclear DNA), 142, 174

New Hope Fertility Center, 176

Newsweek, 102

Nightlight Christian Adoptions, 163

NIH. *See* National Institutes of Health

Nilsson, Lennart, *99*

nonmedical traits, sought-after, 162–63

nuclear DNA (nDNA), 142, 174

Obama, Barack, 190

Obama, Michelle, 203–4, 205

Oberon, Merle, 13–14

"Octomom" scandal, 184

Okun, Doug, 165–66

O'Neill, Kate, 209

online egg market, 160–61

oocyte cryopreservation. *See* cryopreservation

ooplasmic transplantation, 141–42

opposition: to embryo research, 46–47, 48, 63–67, 72–73, 108; to IVF, 60–61, 65, 107; to MRT, 175–77; to regulation of reproductive technologies, 115; to reproductive technologies, 45–48

opposition to abortion, 55, 106–7

oral contraceptives: Garcia and, 80; Pincus and, 231n25; reproductive decisions and, 106; Rock and, 29

Ott, Maureen, 141–42

ovarian resections, 58

ovarian stimulation, 62, 228n7

ovulation, 17, 50, 121–22. *See also* superovulation

ovulation-induction medications: Clomid, 204, 227n7; multiple births and, 193; Pergonal, 1

Palmer, Raoul, 34

parthenogenesis, 18, 19

partial zona dissection, 140

patients: of Asch, 126; characteristics of, 6, 88–89, 93, 94, 203–8; of Coney, 89; consumerist movement among, 116; consumer protection for, 116–19; of Daniell, 51–52; diagnoses of, 94; of Jones Clinic, 61–62, 75–76; of Marrs, 77; of Melbourne IVF team, 75; new categories of, 156–57; of Penn clinic, 81–82, 88; of Pennsylvania Hospital, 82, 88–89; race and ethnicity of, 203–8; of Rock, 12, 25–26; selection of clinics by, 93–96; of Steptoe, 37, 75; vulnerability of, 92

Patton, Mary, 51–52

Paulson, Richard, 132, 200

Pauly, Philip, 18

pelvic inflammatory disease, 73, 153–54

Penn IVF program, 1, 78, 80, 81–83, 85, 88, 151

Pennsylvania Hospital, 82, 83–84, 85–86, 87, 88–89

Pergonal, 1

Petrucci, Daniele, 29, 41

PGD (preimplantation genetic diagnosis), 156–57, 173–74, 198

PGS (preimplantation genetic screening), 4, 174, 198

Phipps, William, 133

Pincus, Gregory, 19–20, 29, 36, 239n8

policy development, 7–8

political culture: chasm between parties, 114–15; embryo research and, 48, 108, 143–44; Ethics Advisory Board and, 65; funding for embryo research and, 7, 52, 63–67, 72–73, 115, 143–44; genetic modification and, 175–76; regulatory impasse and, 119–20, 186, 210–12; reproductive technologies and, 106–8. See also congressional hearings; conservatives/conservatism

pregnancy: age and, 199–202; chemical, 40–41; ectopic, 2, 49; IVF as decoupling genetic parenthood and, x–xi; miscarriage of, 241n71; of older women, 158–60; postmenopausal, 157–58; risks and complications of, 166–67; risks of, 158–60; risks of multiple gestation in, 192. See also gestational surrogacy

preimplantation genetic diagnosis (PGD), 156–57, 173–74, 198

preimplantation genetic screening (PGS), 4, 174, 198

private clinics, 90–93, 127–28, 150, 151–53

programs, components of successful, 84–85

pronatalism in US, 43–48, 71, 103–6

Purdy, Jean, 39, 49, 51, 98, 99

Quigley, Martin, 67, 70, 80, 102, 151

race: access to ART and, 4–5, 129–30, 203–8; infertility and, 105

Ramsey, Paul, 46, 47

Ratcliff, J. D., 19, 27–28

Raymond, Janice, 107

Reagan, Ronald, 71–72, 107

Reed, Candice, 74, 75

Reed, Linda, 74, 75

regulation of reproductive technologies: congressional hearings on, 109–13; failure to implement, 185–86; in Great Britain, 108–9; impasse over, 119–20, 210–12; opposition to and disagreement about, 115; by states, 260n5

Reicher, M-Liz, 130

reproduction, control over, 44

reproductive endocrinology/endocrinologists, 3, 53, 84, 151–52

research: in academic medical centers, 151; on infertility in UK, 40; on investigational new drugs, 176; of Marrs, 69–70; patients as subjects of, 76–77; of Rock, 16–17; on stem cells, 188. See also embryo research

RESOLVE, 106, 118, 154, 206

response to reproductive technologies: political culture and, 106–8; pronatalism, 103–6; "test-tube babies," 8, 54–56; "three-parent embryos," 141

Rice, Heather, 167–68

Richards, Sarah Elizabeth, 170, 172

Roberts, Dorothy, 129–30

Robertson, John, 110

Rock, John, 97; access and, 8–9; career of, ix, 6; DeCherney on, 150–51; donor insemination and, 251n76; editorial by, 20; Garcia and, 239n8; IVF and, 11–12, 13–14, 18,

22–24, 25, 37–38; letters to, 11, 12, 13, 27, 28; life and education of, 14; Mastroianni and, 80; media and, 27–28; medical training of, 15–16; Menkin and, 20–22, 29; research program of, 16–17; treatment of patients by, 25–26; United Fruit Company and, 14–15

Roe v. Wade, 48, 55

Rogers, Paul, 54

Rorvik, David, 42

Rose, Molly, 58

Rubin, I. C., 12

Saarinen, Sharon, 142

same-sex couples: donor insemination and, 243n24; gestational surrogacy and, 165–66, 169–70, 195; reproductive services for, 157; right to marry of, 169

Sanchez, Suzanne, 192

SART (Society for Assisted Reproductive Technology), 92, 118, 171, 186

Sauer, Mark, 127, 132, 138–39, 162, 200, 201

Sawyer, Diane, 125

scandals, 121–26, 184–85, 196

Schenk, Samuel Leopold, 18

schizencephaly, 168

Schoolcraft, William, 153

Schroeder, Patricia, 113–14

Schweitzer, Arlette and Christa, 133

science and technology, attitudes toward, 18–19

semen analysis, 3

sex-change surgery, 58

sex of child, choosing, 173

sexually transmitted infections, 12, 73, 153, 159

Shady Grove Fertility, 150

Shettles, Landrum, 29, 37, 42, 50

Shriver, Sargent, 46

single-embryo transfer, elective, 193

skepticism of IVF, 66–67

Snowflakes program, 163

Society for Assisted Reproductive Technology (SART), 92, 118, 171, 186

Sondheimer, Steven, 2, 81, 82, 83, 149

Sorkow, Harvey, 134

Soupart, Pierre, 51, 65, 235n67

Spar, Debora, 152–53, 211

Speirs, Andrew, 99

sperm, commodification of, 193–98. *See also* insemination; intracytoplasmic sperm injection; male infertility

sperm banks, 141, 194

Sperm Bank USA, 194

Speroff, Leon, 19, 233n22

stem cell research, 188

Steptoe, Patrick: at Bourn Hall, 79, 99; Lesley Brown and, 75; career of, 32, 33; death of, 51; R. Edwards and, 30, 32–33, 34, 37, 48–49, 99; laparoscopic surgery of, 34, 39; Medical Research Council and, 39–40; multiple births after IVF and, 191. *See also* Brown, Louise

Stern, William and Elizabeth, 134

stimulated cycles, 62, 228n7

St. Luke's Hospital, Bethlehem, 89

Stone, Sergio, 123–24, 126

storage: of embryos, 161; of frozen eggs, 203

Streeter, George, 22

subzonal insemination, 140

success rates: components of, 84–85; with frozen eggs, 172; of infertility treatment, 154–55; of IVF, 82, 88, 91, 93, 128–29; misrepresentation of, 126; requirements to report, 118–19, 186

Suleman, Nadya, 184

superovulation, 21, 36, 49, 62

surgical skills, 84–85

surrogacy: altruistic, 133, 196–97, 198; attempt to criminalize, 114; availability of, 195; compensation for, 196–98; complexities of, 164–70; expansion of IVF and, 94; first program for, 96; history of, 134–36; international, 197; H. Jones on, 210; markets for, 196; parental rights and, 136–37; same-sex couples and, 195; "traditional," 133–34; transactional behavior in, 196; use of, 133; Utian and, 134–35

Talbert, Luther, 87

Taylor, Janelle, 202

"test-tube babies." *See* Brown, Louise; in vitro fertilization

Thatcher, Margaret, 108

"three-parent embryos," 141, 185

Index

Tomes, Nancy, 116
transfers: of eggs, 254n24; of embryos, 35–36, 168–69, 191, 193. *See also* GIFT
transplants: ooplasmic, 141–42; uterus, 208–10
transvaginal ultra-sound guided egg retrieval, 90
triplets, risk of premature birth with, 192
"triumphal medicine," age of, 24
Trounson, Alan, 70, *99*
Trump, Donald, 190, 207
tubal factor infertility, 12–13, 73–77
tubal insufflation, 12–13
tubal surgeons, 67
Tureck, Richard, 81
Twardowski, Lorraine and Charlie, 83

unexplained infertility, 203–4
United Kingdom (UK). *See* Great Britain
University of California, Irvine, 122, 123, 126
University of California, San Francisco, Center for Reproductive Health, 206, 207–8
University of Oklahoma, 89
University of Pennsylvania IVF program (Penn), 1, 78, 80, 81–83, 85, 88, 151
University of Southern California (USC), 67–69, 151
University of Texas at Houston, 80, 151
University of Washington, 90–91
uterine factor infertility, 209
uterus transplants, 208–10
Utian, Wulf, 96, 134–35, 169, 195

Vanderbilt University, 80, 151
Vande Wiele, Raymond, 42, 50
Van Steirteghem, Andre, 140
Varmus, Harold, 143
Victor, Baron Rothschild, 37–38

Virginia, IVF in. *See* Eastern Virginia Medical School
vitrification of eggs, 170–71

Wallach, Edward, 82, 85, 86, 87, 88, *101*
Walters, Barbara, 205
Warnock Commission, 108–9
Washington Post editorial board, 61
Watkins, Elizabeth Siegel, 45
Watson, James, 46–47
Weill Cornell Medical College, Center for Reproductive Medicine, 152
Wentz, Ann Colston, 70, 80, *100*, 109
Whitehead, Mary Beth, 134
Wild West of ART, US as, 7, 10
Wilkins, Lawson, 57, 58
"womb-like devices," 210
women: age at first birth, 104, 129; anxiety about changing roles of, 107, 156; claims about infertility of, 102–3, 104–5; equality in workplace and, 202. *See also* motherhood; pregnancy
women's health movement, 44–45
Wood, Carl, 40–41, 62, 70
Wyden, Ron, *100*, 108, 115–18
Wyden Act (Fertility Clinic Success Rate and Certification Act of 1992), 118–19, 122, 128, 186

Yale, 80, 151
Yee, Bill, 123
Younger, Benjamin, 92
Younger, Joan, 23, 24–25, 26

Zeringue, Ernest, 162, 163, 194
Zhang, John, 176
ZIFT (zygote intrafallopian transfer), 254n24
zona pellucida, 37